LANDSCAPES ON THE EDGE

New Horizons for Research on Earth's Surface

Committee on Challenges and Opportunities in Earth Surface Processes

Board on Earth Sciences and Resources

Division on Earth and Life Studies

NATIONAL RESEARCH COUNCIL
OF THE NATIONAL ACADEMIES

THE NATIONAL ACADEMIES PRESS
Washington, D.C.
www.nap.edu

THE NATIONAL ACADEMIES PRESS • 500 Fifth Street, N.W. • Washington, DC 20001

NOTICE: The project that is the subject of this report was approved by the Governing Board of the National Research Council, whose members are drawn from the councils of the National Academy of Sciences, the National Academy of Engineering, and the Institute of Medicine. The members of the committee responsible for the report were chosen for their special competences and with regard for appropriate balance.

This study was supported by the National Science Foundation under award No. EAR-0738045. The opinions, findings, and conclusions or recommendations contained in this document are those of the authors and do not necessarily reflect the views of the National Science Foundation.

International Standard Book Number-13: 978-0-309-10424-9 (Book)
International Standard Book Number-10: 0-309-10424-2 (Book)
International Standard Book Number-13: 978-0-309-10425-6 (PDF)
International Standard Book Number-10: 0-309-10425-0 (PDF)
Library of Congress Control Number: 2010921862

Additional copies of this report are available from the National Academies Press, 500 Fifth Street, N.W., Lockbox 285, Washington, DC 20055; (800) 624-6242 or (202) 334-3313 (in the Washington metropolitan area); Internet http://www.nap.edu.

Cover: The cover photographs emphasize interacting physical, chemical, biological, and human processes on Earth's surface over different scales of space and time. The soil horizon with exposed root, plants, and a human dwelling reveals a complex series of biogeochemical interactions in the near-surface. As the Columbia Glacier on Prince William Sound, Alaska, retreats due to a warming climate, newly scoured rock is exposed and chunks of ice melt into the sound. The circular vignette of a "bare Earth" lidar image near Flathead Lake, Montana, signifies the capability of measuring various features of Earth's surface with precision. Photos courtesy of Robert S. Anderson, University of Colorado, Boulder (Columbia Glacier), Peter Sak, Dickinson College (soil horizon), National Center for Airborne Laser Mapping (lidar image). Cover design by Michael Dudzik.

Copyright 2010 by the National Academy of Sciences. All rights reserved.

Printed in the United States of America

THE NATIONAL ACADEMIES
Advisers to the Nation on Science, Engineering, and Medicine

The **National Academy of Sciences** is a private, nonprofit, self-perpetuating society of distinguished scholars engaged in scientific and engineering research, dedicated to the furtherance of science and technology and to their use for the general welfare. Upon the authority of the charter granted to it by the Congress in 1863, the Academy has a mandate that requires it to advise the federal government on scientific and technical matters. Dr. Ralph J. Cicerone is president of the National Academy of Sciences.

The **National Academy of Engineering** was established in 1964, under the charter of the National Academy of Sciences, as a parallel organization of outstanding engineers. It is autonomous in its administration and in the selection of its members, sharing with the National Academy of Sciences the responsibility for advising the federal government. The National Academy of Engineering also sponsors engineering programs aimed at meeting national needs, encourages education and research, and recognizes the superior achievements of engineers. Dr. Charles M. Vest is president of the National Academy of Engineering.

The **Institute of Medicine** was established in 1970 by the National Academy of Sciences to secure the services of eminent members of appropriate professions in the examination of policy matters pertaining to the health of the public. The Institute acts under the responsibility given to the National Academy of Sciences by its congressional charter to be an adviser to the federal government and, upon its own initiative, to identify issues of medical care, research, and education. Dr. Harvey V. Fineberg is president of the Institute of Medicine.

The **National Research Council** was organized by the National Academy of Sciences in 1916 to associate the broad community of science and technology with the Academy's purposes of furthering knowledge and advising the federal government. Functioning in accordance with general policies determined by the Academy, the Council has become the principal operating agency of both the National Academy of Sciences and the National Academy of Engineering in providing services to the government, the public, and the scientific and engineering communities. The Council is administered jointly by both Academies and the Institute of Medicine. Dr. Ralph J. Cicerone and Dr. Charles M. Vest are chair and vice chair, respectively, of the National Research Council.

www.national-academies.org

COMMITTEE ON CHALLENGES AND OPPORTUNITIES IN
EARTH SURFACE PROCESSES

DOROTHY J. MERRITTS, *Chair*, Franklin & Marshall College, Lancaster, Pennsylvania
LINDA K. BLUM, University of Virginia, Charlottesville
SUSAN L. BRANTLEY, The Pennsylvania State University, University Park
ANNE CHIN, University of Oregon, Eugene[1]
WILLIAM E. DIETRICH, University of California, Berkeley
THOMAS DUNNE, University of California, Santa Barbara
TODD A. EHLERS, University of Tübingen, Germany[2]
RONG FU, University of Texas, Austin[3]
CHRISTOPHER PAOLA, University of Minnesota, Minneapolis
KELIN X. WHIPPLE, Arizona State University, Tempe

National Research Council Staff

ELIZABETH A. EIDE, Study Director
JARED P. ENO, Research Associate (until July 1, 2009)
COURTNEY R. GIBBS, Program Associate
NICHOLAS D. ROGERS, Research Associate (from July 1, 2009)

---

[1] Until August 2008, Texas A&M University, College Station
[2] Prior to September 2009, University of Michigan, Ann Arbor
[3] Prior to August 2008, Georgia Institute of Technology, Atlanta

BOARD ON EARTH SCIENCES AND RESOURCES

CORALE L. BRIERLEY, *Chair*, Brierley Consultancy, LLC, Highlands Ranch, Colorado
KEITH C. CLARKE, University of California, Santa Barbara
DAVID J. COWEN, University of South Carolina, Columbia
WILLIAM E. DIETRICH, University of California, Berkeley
ROGER M. DOWNS, The Pennsylvania State University, University Park
JEFF DOZIER, University of California, Santa Barbara
KATHERINE H. FREEMAN, The Pennsylvania State University, University Park
WILLIAM L. GRAF, University of South Carolina, Columbia
RUSSELL J. HEMLEY, Carnegie Institution of Washington, Washington, D.C.
MURRAY W. HITZMAN, Colorado School of Mines, Golden
EDWARD KAVAZANJIAN, JR., Arizona State University, Tempe
LOUISE H. KELLOGG, University of California, Davis
ROBERT B. McMASTER, University of Minnesota, Minneapolis
CLAUDIA INÉS MORA, Los Alamos National Laboratory, New Mexico
BRIJ M. MOUDGIL, University of Florida, Gainesville
CLAYTON R. NICHOLS, Idaho National Engineering and Environmental Laboratory (retired), Ocean Park, Washington
JOAQUIN RUIZ, University of Arizona, Tucson
PETER M. SHEARER, University of California, San Diego
REGINAL SPILLER, Allied Energy, Houston, Texas
RUSSELL E. STANDS-OVER-BULL, Anadarko Petroleum Corporation, Denver, Colorado
TERRY C. WALLACE, JR., Los Alamos National Laboratory, New Mexico
HERMAN B. ZIMMERMAN, National Science Foundation (retired), Portland, Oregon

National Research Council Staff

ANTHONY R. DE SOUZA, Director
ELIZABETH A. EIDE, Senior Program Officer
DAVID A. FEARY, Senior Program Officer
ANNE M. LINN, Senior Program Officer
SAMMANTHA L. MAGSINO, Program Officer
MARK D. LANGE, Associate Program Officer
LEA A. SHANLEY, Postdoctoral Fellow

JENNIFER T. ESTEP, Financial and Administrative Associate
NICHOLAS D. ROGERS, Financial and Research Associate
COURTNEY R. GIBBS, Program Associate
JASON R. ORTEGO, Research Associate
ERIC J. EDKIN, Senior Program Assistant
TONYA FONG YEE, Senior Program Assistant

Preface

In 2008, some 60 early-career Earth scientists met to explore the dynamic interactions of life and its landscape—in essence the interplay between biological and physical Earth surface processes. Topics ranged in scale from sand grains to continents, and enthusiasm for all ran high. This National Science Foundation (NSF)-sponsored Meeting of Young Researchers in Earth Science (MYRES) readily achieved its aim of interdisciplinary community building. New teams of collaborators forged during the workshop continued to meet and communicate throughout the year, generating bold ideas for research that bridged multiple fields. Many of their ideas were so unconstrained by conventional disciplinary boundaries that the researchers were hard-pressed to identify the appropriate agencies and programs that might fund such investigations. Indeed, what made the meeting so successful was the rare opportunity for scientists from different fields, who might not go to the same sessions at scientific meetings or even perhaps the same meetings, to interact and explore topics with fresh, multifaceted perspectives. These perspectives came from individual expertise honed in climate science, ecology, geochemistry, geography, geomorphology, hydrology, soil science, and other disciplines, but the power of the new ideas drew from the merging of perspectives to scrutinize compelling scientific questions.

The Committee on Challenges and Opportunities in Earth Surface Processes, also comprising scientists from a variety of disciplinary backgrounds (Appendix A), had a somewhat similar experience that began earlier, in late 2007, and extended over an 18-month period. Convened by the National Research Council at the request of NSF to assess the current state of the field of Earth surface processes, the 10 committee members, aided by community input (Appendix B), began immediately to identify fundamental, overarching scientific questions. What controls, for example, how resilient landscapes are to climate, tectonic, and land-use changes? What can we learn about possible future Earth conditions from Earth's past, as recorded in its landscapes, sediments, and soils? How do ecosystems and landscapes coevolve, and how can landscapes be restored or redesigned in sustainable

PREFACE

ways? The intellectual excitement of these and other ideas presented in this report lies in their integrative nature and potential to bring together scientists from diverse fields who share in common the object of study, the Earth's surface.

The committee identified the opportunities as well as the rate-limiting challenges for making significant research advances in understanding the form, composition, properties, function, and evolution of Earth's surface. Two of the most significant research opportunities are obvious: (1) remarkable new technologies exist for exploring Earth's surface and for dating sediments, landforms, and soils; and (2) dramatic changes have occurred on many parts of the planet in response to a variety of anthropogenic activities. The most significant technical challenges include a dearth of sites with instrumentation for long-term (decadal or more) monitoring of basic Earth surface characteristics and processes, such as soil moisture and temperature, eolian dust transport, sediment transport, or water chemistry and stream flow.

The most fundamental intellectual challenges are that the state of the science of Earth surface processes has been reliant primarily upon empirical approaches, and that the interdisciplinary communities needed to advance the science are nascent. Although empirical approaches are necessary and have produced some notable achievements, quantitative (mechanistic) theories and approaches are needed to describe the processes that shape and alter the composition of Earth's surface. With combined approaches and disciplinary perspectives, we can make more confident predictions of how landscapes respond to interacting changes. If climate change brings, say, an increase in permafrost melting, will thawed soils deliver amplified sediment and nutrient pulses to streams and coastal water bodies? If so, what feedbacks might occur as biogeochemical fluxes are altered?

The landscape, its evolution, and its role as the arena for life and human activity—although intricate—are becoming comprehensible and even predictable as we work across intellectual barriers that have fragmented landscape research in the past. We imagine how the study of Earth surface processes could be transformed if events such as the MYRES workshop were commonplace, and if the climate scientists, ecologists, geographers, geomorphologists, geochemists, and others who study Earth surface processes were to have increasing opportunities to meet, share their disciplinary strengths, generate new ideas, and investigate compelling scientific problems. Our current situation—one of concern about landscapes on the edge of potentially detrimental and irreversible change—could shift to one of optimism about science and society "on the edge" of new understanding of landscape change and processes at Earth's surface.

This vision goes beyond merely the hope for more frequent and sustained interactions among scientists in presently fragmented disciplines that focus upon different aspects of landscape processes. Acquisition and management of multiple types of data are needed, as are the development and support of new tools for exploring Earth's surface and establishing the nature of its past environmental states.

Preface

This report is intended to inspire scientists from both the natural and the social sciences to establish broad-reaching research agendas to solve some of the key research questions about Earth surface processes operating in the past, present, or future. The report is equally supportive of the basic, discipline-specific research sponsored by NSF and practiced at multiple federal agencies. Discipline-specific research and expertise are the foundation for productive interdisciplinary endeavors, and both types of research are necessary parts of a national research support structure.

This is an auspicious time for the study of Earth surface processes, with a rapidly growing base of new scientists, a wealth of relatively unexplored scientific questions, and a new cache of powerful investigative tools. The timing also is promising in that NSF's Geovision document is directed partly toward new, interdisciplinary initiatives. We hope that the present report can be synchronized with some of the goals promoted by NSF in the Geovision document. This report also is timely, given the changes in Earth's surface as a result of human and natural impacts. The committee is optimistic that the ideas and content of this report will promote a collective effort to address a number of societally relevant and scientifically compelling challenges in the field of Earth surface processes. New and enhanced interdisciplinary partnerships, coordinated monitoring and observation of the Earth's near-surface systems and new scientific theories for processes of weathering, erosion, transport, and deposition of Earth surface materials could transform the field, ultimately leading to much greater capacity to predict how Earth's surface will change in the future.

Working with the scientists on this committee and hearing from colleagues who represented numerous agencies and universities throughout our deliberations provided many days of intellectual enjoyment. I thank the committee for its dedication, stamina, and hard work. The committee is grateful to Elizabeth Eide, the project study director, for her extraordinary abilities in many aspects, but in particular for her capacity to meld numerous comments and feedback into meaningful, thorough compilations that guided us as we developed the report. The committee also benefited from the dedication and excellence of research associates Jared Eno and Nicholas Rogers, and program associate Courtney Gibbs. As chair, I offer a personal note of thanks to all of these fine individuals.

Dorothy J. Merritts
Chair

Acknowledgments

In addition to its own expertise, the committee relied upon input from numerous external professionals with extensive experience in Earth surface processes. These individuals provided written and oral contributions, which were very important to the committee in formulating this report. We would like to express our appreciation to the many highly qualified individuals who provided testimony, data, and advice during the course of the study and aided the committee in reaching diverse corners of the broad community of researchers involved in Earth surface processes; in particular, the committee would like to thank: Teofilo Abrajano, Robert Anderson, Ramon Arrowsmith, Rafael Bras, Oliver Chadwick, Terry Chapin, Michael Church, Louis Derry, Martin Doyle, Tom Drake, Michael Ellis, Jon Foley, Christian France-Lanord, Joseph Galewsky, Arthur Goldstein, Will Graf, Neal Iverson, Richard Iverson, Matthew Larsen, Eli Lazarus, Randy McBride, Gregory Okin, Aaron Packman, Frank Pazzaglia, Denise Reed, Linda Rowan, Fred Scatena, John Shroder, Robert Stallard, Brad Werner, James Whitcomb, and Izzy Zietz.

This report has been reviewed in draft form by individuals chosen for their diverse perspectives and technical expertise, in accordance with procedures approved by the National Research Council's Report Review Committee. The purpose of this independent review is to provide candid and critical comments that will assist the institution in making its published report as sound as possible and to ensure that the report meets institutional standards for objectivity, evidence, and responsiveness to the study charge. The review comments and draft manuscript remain confidential to protect the integrity of the deliberative process. We wish to thank the following individuals for their participation in the review of this report:

Suzanne Anderson, University of Colorado, Boulder
Christopher Beaumont, Dalhousie University, Halifax, Nova Scotia, Canada
Sergio Fagherazzi, Boston University, Massachusetts
Kenneth A. Farley, California Insitute of Technology, Pasadena

ACKNOWLEDGMENTS

Scott Fendorf, Stanford University, California
Jon Harbor, Purdue University, West Lafayette, Indiana
Zhengyu Liu, University of Wisconsin, Madison
Glen M. MacDonald, University of California, Los Angeles
Richard Marston, Kansas State University, Manhattan
David Mohrig, The University of Texas at Austin
Sujith Ravi, University of Arizona, Tucson

Although the reviewers listed above have provided many constructive comments and suggestions, they were not asked to endorse the conclusions or recommendations nor did they see the final draft of the report before its release. The review of this report was overseen by William E. Easterling, III, The Pennsylvania State University, who was appointed by the National Research Council and was responsible for making certain that an independent examination of this report was carried out in accordance with institutional procedures and that all review comments were carefully considered. Responsibility for the final content of this report rests entirely with the authoring committee and the institution.

Contents

SUMMARY   1

1   THE IMPORTANCE OF EARTH SURFACE PROCESSES   13
    1.1   Introduction, 13
    1.2   Examples of Interconnected Earth Surface Processes, 16
    1.3   New Technologies: Monitoring Earth Surface Processes at High Resolution in Space and Time, 24
    1.4   Study Considerations and Report Structure, 28
    1.5   Closing Remarks, 32

2   GRAND CHALLENGES IN EARTH SURFACE PROCESSES   35
    2.1   What Does Our Planet's Past Tell Us About Its Future?, 35
    2.2   How Do Geopatterns on Earth's Surface Arise and What Do They Tell Us About Processes?, 42
    2.3   How Do Landscapes Influence and Record Climate and Tectonics?, 50
    2.4   How Does the Biogeochemical Reactor of the Earth's Surface Respond to and Shape Landscapes from Local to Global Scales?, 61
    2.5   What Are the Transport Laws That Govern the Evolution of the Earth's Surface?, 68
    2.6   How Do Ecosystems and Landscapes Coevolve?, 78
    2.7   What Controls Landscape Resilience to Change?, 83
    2.8   How Will Earth's Surface Evolve in the "Anthropocene"?, 93
    2.9   How Can Earth Surface Science Contribute Toward a Sustainable Earth Surface?, 102

CONTENTS

3 FOUR HIGH-PRIORITY RESEARCH INITIATIVES IN EARTH SURFACE PROCESSES   109
    3.1 Interacting Landscapes and Climate, 109
    3.2 Quantitative Reconstruction of Landscape Dynamics Across Time Scales, 112
    3.3 The Coevolution of Ecosystems and Landscapes, 113
    3.4 The Future of Landscapes in the "Anthropocene," 114
    3.5 Summary, 116

4 MECHANISMS FOR DEVELOPING INITIATIVES AND SUSTAINING GROWTH IN EARTH SURFACE PROCESSES   119
    4.1 Federal Research Framework, 119
    4.2 Other Partnerships, 122
    4.3 Data Collection and Distribution, Modeling, Tool Development, and Community Research Facilities and Sites, 123
    4.4 Developing the High-Priority Research Initiatives, 128

REFERENCES   133

APPENDIXES

A Biographical Sketches of Committee Members and Staff   139
B Community Input   145
C Observing and Measuring Earth Surface Processes   149
D Achievements in Earth Surface Processes   155
E List of Acronyms   161

Summary

S.1 INVESTIGATIONS OF EARTH'S SURFACE

Earth's surface is a dynamic interface across which the atmosphere, water, biota, and tectonics interact to transform rock into landscapes with distinctive features crucial to the function and existence of water resources, natural hazards, climate, biogeochemical cycles, and life. Interacting physical, chemical, biotic, and human processes—"Earth surface processes"—alter and reshape Earth's surface on spatial scales that range from those of atomic particles to continents and over time scales that operate from nanoseconds to millions of years. The study of Earth surface processes and the landscapes they create is rich with open questions and opportunities to make fundamental scientific advances and to understand and predict the interactions, causes, and effects of these processes. Scientists who study Earth's "surface processes" have a distinctive and novel ability to contribute to understanding how Earth's surface changes with time and resolving important environmental challenges that may arise from these changes.

Research in Earth surface processes has grown significantly in the last two decades, in response largely to two factors. First, scientists, policy makers, and the public have become increasingly aware of the impact of human activity and climate change on Earth's surface. The changes to Earth's surface affected by natural events and by humans, notably through land use, have altered the physical, chemical, and biological integrity of soils, mountains, prairies, rivers, coasts, and watersheds. Thus, society has heightened its demand for scientific guidance in making decisions concerning the future of Earth's surface. Second, development of new analytical and computing tools has markedly increased our ability to examine Earth's surface at high spatial and temporal resolution and to develop models that can help to understand the speed and magnitude over which surface processes interact and affect changes.

Recognizing the growing importance of understanding Earth's surface and the processes that shape it, the National Science Foundation (NSF) requested the National Research

> **BOX S.1**
> **Statement of Task**
>
> Earth's surface is a dynamic interface where physical, chemical, biological, and human processes cause and are affected by forcings in the Earth system. This impact-feedback loop occurs over a wide range of temporal and spatial scales. It binds the Earth's surface to a host of scientific and societal issues, and within this context the committee will:
>
> 1. Assess the state of the art of the disciplinary field of Earth surface processes and the fundamental research questions in the field;
> 2. Identify rate-limiting challenges or opportunities for making significant advances in the field; and
> 3. Identify the necessary intellectual collaborations and high-priority needs to meet these challenges.

Council (NRC) to establish a committee to address challenges and opportunities in Earth surface processes (Box S.1).

In response to its charge, the committee has identified nine overarching scientific challenges for increasing our understanding of Earth's surface processes and has established four high-priority research initiatives drawn from these challenges. The four research initiatives emphasize the dynamic interactions among the various processes operating on Earth's surface and require fundamentally new interdisciplinary research approaches to develop and support them. Designed to transform and strengthen the field of Earth surface processes, the initiatives represent promising pathways to meet urgent demands for scientific guidance on issues related to planning, mitigation, and response to changes in Earth's surface now and in the future. Some of the key intellectual and technical barriers to advance these research initiatives also are identified, as are strategies to overcome these barriers. Because the committee comprised primarily Earth scientists, the focus of the report is on the terrestrial Earth surface.

S.2 GRAND CHALLENGES IN EARTH SURFACE PROCESSES

Each of the nine grand research challenges identified by the committee is presented in the form of a primary research question with associated opportunities for further investigation. The grand challenges are representative of exemplary, interdisciplinary research areas in Earth surface processes deemed by the committee as poised for significant advances. The list is not intended to be a comprehensive list of all important research activities in Earth surface science:

1. *What Does Our Planet's Past Tell Us About Its Future?* This challenge highlights the extent to which Earth's surface system records its own evolution, how that record can be tapped to understand changes in the surface environment through time, and thus how Earth's surface may change in the future. Some of the key opportunities within this challenge include (1) accelerating the ability to reconstruct past Earth surface history quantitatively through the examination, for example, of detailed sediment or ice core records and (2) measuring the rate of surface change through application of surface-dating methods.
2. *How Do Geopatterns on Earth's Surface Arise and What Do They Tell Us About Processes?* Earth surface processes create myriad spatial patterns at all scales—from microscopic fabrics observed in soils, to the repeated patterns of sand dunes and barrier islands. An unmistakable human footprint also is evident on many landscapes. These geopatterns often are self-organized—emerging spontaneously from local interactions rather than being imposed by some outside influence—and dynamic, in that they develop and evolve over time. They also are resilient; unstable patterns do not persist. Landscape geopatterns provide a template for understanding a broad spectrum of processes on the land surface. This understanding can be used to improve predictions of the response of the Earth's surface to natural and human-induced changes.
3. *How Do Landscapes Influence and Record Climate and Tectonics?* One of the major advances in the Earth sciences of the last two decades is recognition of both broad and more subtle connections between climate and tectonic systems. Some of the most intriguing research questions in the interaction of landscapes, climate, and tectonics center on the relative sensitivity and rates of the numerous feedback mechanisms among climate, topography, ecosystems, physical and chemical denudation, sedimentary deposition, and the deformation of rocks in active mountain belts. Four research opportunities emerge as particularly promising to advance understanding of these linkages: (1) quantification of the role of climate in surface processes; (2) influence of mountain building and surface processes on climate; (3) sedimentation and mountain building; and (4) interactions of surface processes, climate, tectonics, and mantle dynamics.
4. *How Does the Biogeochemical Reactor of the Earth's Surface Respond to and Shape Landscapes from Local to Global Scales?* Chemical erosion and weathering of bedrock and soil are major factors in Earth surface processes because of their potential effects on climate, the chemistry of groundwater and rivers, the strength of rocks, the erodibility of landscapes, the availability of nutrients in soils, the fate of anthropogenic contaminants, and the properties of ecosystems. Nonetheless, the breakdown of bedrock and soil is among the least understood of the important geological processes, perhaps most significantly with respect to the manner in which biogeochemical reactions affect carbon cycling. Understanding the biogeochemical

cycling of elements needed by biota and essential to Earth's climate in the present and the past will allow better understanding, for example, of the magnitude of the effects of human-induced land-cover change on the balance of biogeochemical cycles. Advances in technologies to monitor and analyze biogeochemical changes in shallow subsurface geophysics, in geobiology, in nanogeoscience, and in rock mechanics will help to move research in this area from correlation to explanation and from observation to prediction.

5. *What Are the Transport Laws That Govern the Evolution of the Earth's Surface?* A mechanistic (quantitative and process-based) understanding of links among climate, hydrology, geology, biota, land use, topography, and rates of erosion and deposition is a fundamental goal of Earth surface process research. To tackle this challenge we need to discover, quantify, test, and apply mathematical laws that define the rates of processes shaping Earth's surface. Significant progress has been made recently in developing and applying laws for the mass transport of soil on hillslopes and river incision into bedrock. However, we still lack transport laws for such fundamental processes as landsliding, overland flow erosion, glacial erosion, chemical erosion, and transport and deposition of mud. The breakdown of bedrock into erodible debris, the first step in hillslope erosion, also is poorly understood. The study of landscape history is essential for testing landscape evolution theories.

6. *How Do Ecosystems and Landscapes Coevolve?* Life—through digesting, dilating, exhaling, decaying, pushing, and weaving—strongly influences the form and pace of surface erosion and modulates biogeochemical cycling, with simultaneous effects on climate, hydrology, erosion, and topography. Recent developments indicate that we are in a position to make significant advances in understanding the coevolution of landscapes, life, and ecosystems. New opportunities to investigate these interactions are found in the emerging fields of geobiology, ecohydrology, and ecogeomorphology. Coordinated efforts that explicitly address linkages among biota, Earth surface processes, and landforms are under way at field observatories. However, a need exists for greater mechanistic understanding of life-landscape interactions in order to make modeling predictions and perform experiments to explore and discover the causes, effects, rates, and magnitudes of these interactions.

7. *What Controls Landscape Resilience to Change?* The shapes of landforms and their rates of evolution fluctuate within ranges, reflecting the stochastic nature of the processes that drive their operation. However, when conditions change with sufficient magnitude and duration, landscapes may become altered beyond the range within which they can recover. The driving agents have then exceeded the resilience of the landscape. Some areas of the Earth's surface are more vulnerable than others to changes in state. Polar, glacial, and periglacial regions, for example, currently are nearing, or are in, a change of state predicted to continue with global warming. This

research challenge highlights the state of scientific knowledge related to rapid and abrupt change on Earth's surface and to the factors and processes that control the resilience of landscapes to change. The science goals include identifying thresholds of change, understanding the environmental processes most vulnerable to change, understanding the mechanisms that make some landscapes resilient to change, and investigating options for mitigating or even reversing the effects of change. Studying the impact of abrupt changes on Earth's surface requires examining both geologic and recent records of their occurrence.

8. *How Will Earth's Surface Evolve in the "Anthropocene"?* The environmental impacts of human activities are pervasive. The term "Anthropocene" has emerged in the scientific literature to suggest the onset of a new geologic era in which humans have become dominant. This overprint of human actions on Earth's surface has made even the identification of "natural" landscapes difficult. Understanding, predicting, and adjusting to changing landscapes increasingly altered by humans constitute pressing challenges that fall squarely within the purview of Earth surface science. Mechanistic models that account explicitly for human-landscape interactions are needed, especially for adaptive management and for assisting decision making in the face of change. Science is far from developing a general theory of coupled human-natural systems, even though such a theory may offer the potential to slow or reverse environmental degradation. Because such a theory would include knowledge of societal perceptions of environmental impacts and the ability and willingness of societies to react to these changes, much focused inductive and empirical work is still required to investigate these interacting processes in a range of Earth surface environments and human societies.

9. *How Can Earth Surface Science Contribute Toward a Sustainable Earth Surface?* With increasing scientific understanding of the causes and cumulative, long-term effects of human-induced changes to Earth's surface, a consensus has emerged that at least some of these disrupted or degraded landscapes can and should be "restored" or "redesigned". Specific research challenges arise about the size and geometry of redesigned rivers, tidal creeks and inlets, deltas, and beaches and how they are controlled by the expenditure of flow energy and by sediment supplies. Landscape restoration is a complex, high-priority goal for many researchers, practitioners, policy makers, and the public. These different communities have only recently begun to examine together the successes and limitations of past restoration efforts using new data from specific regions of Earth's surface. Earth-surface scientists can contribute to these efforts as restoration activities move toward quantitative models and predictions. This research will provide guidance for future decisions regarding both natural and managed landscapes and will be critical for enhancing the goods and services that Earth's surface provides for society.

S.3 FOUR HIGH-PRIORITY RESEARCH INITIATIVES

Overarching elements of the nine grand challenges in Earth surface processes research are synthesized into four major research initiatives. Timely and rich in scientific merit, these high-priority research areas are expected to transform the field of Earth surface processes. Because the initiatives emphasize the interacting physical, chemical, biological, and human processes on Earth's surface, they will require coordinated, sustained, interdisciplinary approaches to develop new intellectual collaborations among scientists, and to generate new scientific approaches, tools, and models. The emerging science of Earth surface processes remains challenged by intellectual and disciplinary barriers that have fragmented research in this area in the past and by the fact that the science often has relied on fairly simple, descriptive approaches. A variety of mechanisms existing within NSF could provide the research support necessary to help develop the initiatives. The scientific objectives of each initiative are described below, together with specific examples of implementation mechanisms to establish and support these initiatives.

Interacting Landscapes and Climate

A major research initiative in the area of climate-landscape interactions (Box S.2) would develop a quantitative understanding of climatic controls on Earth surface processes and the influence of landscape on climate over time scales that range from individual storm events to the evolution of landscapes. This initiative has the potential to transform our understanding both of the role of climate in changing the Earth's surface and of the feedbacks between surface change and climate.

The primary science objectives for this initiative include the following:

- Development of theory for the interactions among topography, land cover, and global, regional, and local climate that determine the biogeochemically and geomorphically significant attributes of climate
- Development of geomorphic transport laws that explicitly account for climate and incorporate interactions with biota, including theories for river and glacier incision; production, transport, and deposition of sediment; and geochemical processes
- Monitoring, experimentation, and modeling of climatic controls on the weathering of rock and soil and their influence on physical erosion rates and vice versa
- Study of the feedbacks between global and regional climate and (1) the operation of terrestrial carbon reservoirs and (2) the controls on atmospheric dust concentration
- Modeling and monitoring of landscape evolution under diverse and varying climatic conditions; identification of climatic signatures in landscapes; and evaluation of thresholds of landscape response and the limits of resilience

> **BOX S.2**
> **Potential Implementation Mechanisms:**
> **Interacting Landscapes and Climate**
>
> Collaboration between Earth surface scientists and atmospheric scientists is the primary need for developing and advancing this initiative. Improved communication between these communities is essential and could be developed, for example, through the following:
>
> - Workshops, joint field campaigns and summer schools involving climate scientists and Earth surface scientists that address fundamental problems of the interaction of climate and Earth surface processes
> - Collaborative model-building efforts between climate scientists and Earth-surface scientists that may include effects of land cover for regional and microclimate modeling, wind and wave energy, and glacier dynamics
> - Instrumentation advances that join engineers and Earth surface scientists in the development of satellite- and land-based sensors to monitor factors relating to climatic control of Earth surface processes (for example, rainfall, soil temperature, and glacier sliding velocities)

- Development of theories for subglacial hydrology, basal sliding of glaciers, subglacial sediment deformation, and ocean-ice interactions at ice-sheet margins
- Improvement of the coupling between surface processes and existing climate models, explicitly incorporating the effects, feedbacks, and conditions outlined above

Quantitative Reconstruction of Landscape Dynamics Across Time Scales

This major research initiative (Box S.3) is focused on developing quantitative, detailed reconstructions of Earth surface evolution from instants to eons based on information recorded in landscapes and the sedimentary record. A confluence of interest exists between developing an understanding of surface evolution over various time scales and mining the long-term, sedimentary archive for information on the variability of surface dynamics. This archive applies, in particular, to the frequency of rare but potentially catastrophic events. Reconstructed evolution of the Earth's surface will be used (1) to test and develop models that couple tectonics, climate, biota, lithology, and landscape evolution; (2) to constrain the frequencies and causes of rare but important surface events; and (3) to provide baseline information on pre-human landscapes and their response to change as a guide for restoration and management.

> **BOX S.3**
> **Potential Implementation Mechanisms:**
> **Quantitative Reconstruction of Landscape Dynamics Across Time Scales**
>
> The necessary intellectual collaborations for this initiative to succeed are primarily those between researchers studying deep- and surface-Earth processes, from both academic and industry backgrounds. Addressing the objectives in this initiative requires (1) application and development of specific analytical techniques and tools and (2) incorporation of data acquired with these tools in coupled models. Specific mechanisms to foster these collaborations include the following:
>
> - Development of natural deep-time laboratories to focus on reconstruction of Earth surface evolution from short-term to geologic time scales
> - Targeted projects with joint industry and academic participation, organization, and products to apply noncommercially sensitive portions of three-dimensional seismic surveys to key initiative objectives
> - Continued development of cosmogenic, optically stimulated luminescence, and isotopic and low-temperature thermochronological methods, and encouraging use of existing community laboratories and research that apply these tools to new minerals
> - Coordinated community development of fully coupled climate—tectonic, geochemical, ecological—surface process models that engages existing numerical modeling initiatives and organizes interdisciplinary workshops to include atmospheric, ecological, and paleontological disciplines
> - Development and support of shared, community experimental laboratory facilities for landscape research across time scales and testing of models in a controlled environment

The primary science objectives for this initiative include the following:

- Improving methods for quantitative reconstruction of past Earth surface states from the record of landforms, paleobotany, geochemistry, paleo-soils, and sedimentary deposits
- Developing detailed paleoclimate, tectonic, and sedimentary records of abrupt changes in Earth surface processes and of landscape resilience over long time scales to understand the tolerable limits of stochastic variability within different geomorphic systems
- Developing and testing quantitative predictive models for the Earth surface system across time scales with focus areas that include using realistic crust and mantle rheologies; coupling to mantle convection; modeling glacial erosion and transport; and coupling to biogeochemical cycles

- Developing and improving technical capabilities and data collection in near-surface geophysical methods to image and measure Earth's near-surface structure and physical properties in three dimensions

Coevolution of Ecosystems and Landscapes

With new ways of measuring how the living and the nonliving surfaces coorganize and an increasing ability to link biotic processes and landscape evolution, the opportunity exists to forge a new understanding of the coevolution of ecosystems and landscapes (Box S.4) and to address pressing problems of future environmental change. This initiative could lead to the transfer of concepts and theories on physical systems to ecological research and to the use of ecological principles to guide coupled ecosystem and landscape modeling. A primary goal is to build the capability to predict future coupled ecosystem and landscape states under varying climate and land-use conditions.

BOX S.4
Potential Implementation Mechanisms:
Coevolution of Ecosystems and Landscapes

This initiative emphasizes work at the interface of Earth and biological sciences. The ultimate goal is to create opportunities for discoveries that are equally advanced in the fields of ecology and Earth surface processes and are obtainable only with strong interdisciplinary interactions. Specific mechanisms to develop these collaborations include the following:

- Establishing working groups that organize regular meetings to focus on research at the interface of ecosystems and landscape processes and evolution and on opportunities for interdisciplinary teaching and formulating joint research plans—special sessions on ecosystems and landscapes at national meetings of many organizations are a start
- A community-level modeling program for ecologists and Earth scientists to collaborate on models for short-term forecasts and long time-scale predictions of the coevolution of ecosystems and landscapes
- Joint field campaigns conducted by ecologists and Earth surface scientists, including climate scientists, geomorphologists, and hydrologists, to help identify and quantify underlying mechanisms that link biota, ecosystems, and Earth surface processes
- Employing the network of observatory sites to explore ecological and Earth surface processes
- Co-development of instrumentation, geochemical, and geochronological tools that could facilitate significant advances

The primary science objectives for this initiative include the following:

- Improvement of theory and observations that relate spatial patterns and dynamics of biota to landscape setting (topography, hydrology, and geology) for given climatic conditions
- Development of models that incorporate both the geomorphic transport laws and the requisite biogeochemical equations to account mechanistically for the role of biota
- Development of landscape evolution theory that includes the effects of biota (and its possible coevolution with landscape)
- Development of models to predict the coevolution of climate, biota, and landscape processes under a scenario of increased greenhouse gases
- Development of observations and models for the interaction of biota with stream channel and floodplain morphology and dynamics

Future of Landscapes in the "Anthropocene"

Substantial advances have been made in understanding the range and extent of human impacts on Earth surface systems. These advances, coupled with technological breakthroughs, present opportunities to provide answers to a fundamental and urgent question: How can we understand, predict, and respond to rapidly changing landscapes that are increasingly altered by humans? The overarching goal of this initiative is to transform our understanding of integrated human-landscape systems and our ability to predict how they might evolve in the future (Box S.5).

The primary science objectives for this initiative include the following:

- Improved understanding of the long-term legacies of human impacts on landscapes and quantification of current rates of impacts, especially in environments that are sensitive to global climate change
- Development of mechanistic models linking multiple and cumulative effects of human activity
- Development of integrated models of the complex interactions within human-dominated landscapes, incorporating decision making and human behavior
- Greater understanding and predictive capacity for coupled human-landscape dynamics
- Capacity building to anticipate and guide options for mitigating, reversing, and adapting to human-caused landscape change
- Coordinated collection and database management of sociological and geographical information on land use for incorporation into quantitative models

> **BOX S.5**
> **Potential Implementation Mechanisms:**
> **Future of Landscapes in the "Anthropocene"**
>
> Developing and advancing this initiative require new collaborations among Earth surface scientists and a range of social and behavioral scientists—including economists, political scientists, sociologists, and human geographers—to incorporate decision making and human perception and behavior in quantitative models. Collaboration with geospatial scientists is further needed to integrate geospatial technologies into modeling and field-based efforts, as is collaboration with engineers and applied practitioners to use manipulated landscapes for experimentation. Increased collaboration among Earth surface scientists, ecologists, and climate scientists, among others, is also needed to investigate the interacting processes within human-dominated landscapes. These collaborations could be advanced through:
>
> - Workshops for Earth surface scientists and social scientists to build integrated community approaches, research questions, methodologies, scales of inquiry, and theories for human-landscape systems
> - Workshops for geospatial scientists and Earth surface scientists to examine the integration of geospatial technologies with experimental, modeling, and field approaches and to process and synthesize remote-sensing data
> - Development of community field and modeling centers to acquire the data necessary for new integrative and predictive models that involve multiple stressors within human-dominated landscapes, including social processes that influence those interactions
> - Focused field studies in sensitive environments vulnerable to anthropogenic and climate change, including coastal and urban areas, mountain and polar environments, and arid and semiarid areas—existing environmental observatories could be employed
> - Collaborative research using engineered landscapes and restoration and redesign projects. These relatively controlled research conditions can improve fundamental knowledge of processes relevant to a range of environments and problems. This collaboration could involve engineers and applied practitioners working with Earth surface scientists
> - Use of restoration and redesign projects to test hypotheses about how to build self-maintaining ecogeomorphic systems

S.4 CONCLUDING REMARKS

Earth's surface is the only habitat available to humans, and understanding the processes by which that habitat has been created and the ways in which it changes is important to determining the causes of environmental degradation, to restoring what is degraded, and to guiding decisions toward a sustainable future. We have the technological ability to monitor closely the response of landscapes to climate change and human activities, as well as to interactions with other Earth surface processes. The need to model all of these interactions in a predictive fashion is clear. To develop this capability, fundamental research is needed to understand the process-based linkages among life, climate, tectonics, human activity, and landscapes. Environmental restoration and design requires thoughtful consideration of Earth surface processes, landscape history, and the interactions between human activities and surface processes. This area is of great consequence for contributions from scientists in the integrative, emerging field of Earth surface processes that lies at the intersection of natural science disciplines in Earth, life, atmospheric, ocean, and social sciences. With the rise of new scientific questions relating various components of the Earth system, new opportunities and tools for research, rapid growth of human population, and unprecedented changes in biota, land cover, process rates, and global climate, an appraisal of the study of Earth surface processes is timely and crucial.

CHAPTER ONE

The Importance of Earth Surface Processes

1.1 INTRODUCTION

Earth's surface is the arena for most life and all human activity, yet what lies beneath our feet is as mysterious as it is familiar. Earth scientists or not, we recognize hills, mountains, glaciers, deserts, rivers, wetlands, and shorelines. If a good deal of rain falls, floods may occur; if a storm strikes the coast, the beach may erode; if we are careless with our soil, we may damage or even lose it. These ideas are well known, but with just a few questions we arrive at the edge of our knowledge and face gaps that matter to our safety, our food and water security, the infrastructure of roads and river navigation, and the survival and diversity of ecosystems and services they provide.

Any familiar landscape illustrates the point (Figure 1.1). Start with a stream channel and ask a series of simple questions: What controls its size, pattern, and magnitude of flooding? What plants and animals live in and along this stream, and how do biological processes—including human activities—affect the downstream flow of nutrients and water? Next, look about and wonder how this stream relates to its valley and the surrounding hillslopes. How did these landforms arise, and how are they related to one another? Why are hillslopes usually mantled with soil, and why is that soil so much richer and more complex than simple ground bedrock? In addition to landforms and their mantling soil, landscapes host a set of interconnected ecosystems, both visible and microscopic. How have these ecosystems shaped and been shaped by Earth's surface? How is the flow of nutrients that nourishes ecosystems connected to the landscape? Finally, if we take the longest view, our stream is part of a network that forms a kind of continental circulatory system, carrying water, sediment, nutrients, and biota from high ground to low-lying coastlines. How did this system come to be, how long has it existed, and how is it related to climate (modern and past) or to the tectonic forces that shape continents? How will it behave in the future, and how do human activities influence that behavior?

FIGURE 1.1 Landscapes at Earth's surface host a suite of interconnected landforms and processes that can remain stable for long periods of time and can also respond rapidly to changes in climate or land use. In this view of a recently deglaciated valley in the Juneau Icefield, Alaska, surface features comprise hillslopes, rock falls and slides, glaciers (in the far distance, upper right corner of the image), alluvial fans, streams, wetlands, and biota. Integral processes less visible than the landforms and land cover include weathering, soil formation, climate, surface and groundwater flow, nutrient fluxes, and tectonics. SOURCE: Photograph courtesy of Dorothy Merritts, Franklin and Marshall College, Lancaster, Pennsylvania.

Other than a basic goal of explaining the form, composition, and evolution of landscapes, why might questions about Earth surface processes near a stream, or similar types of questions posed along a coastline or in a fragile arctic landscape, matter? At present, we are unable to make confident, process-oriented predictions of how landscapes respond to change. If climate change brings, for example, an increase in rainfall, will soils deliver more or fewer nutrients to groundwater and streams? If humans remove river dams and release the sediment stored behind them, as well as the nutrients and pollutants bound to the sediments, how will downstream fish habitats, estuaries, and coastal marshes be affected? Will the extra sediment stop the retreat of receding beaches, or will the sediment wash out to sea? Because of these and other such critical questions, society has become concerned about landscapes "on the edge" of potentially detrimental and irreversible change and has heightened its demand for scientific guidance in making decisions concerning the future of Earth's surface in light of these changes.

Spurred by growing recognition of the importance and relevance of research in these areas, the National Science Foundation (NSF) requested that the National Research Council (NRC) convene a committee to address challenges and opportunities in Earth surface processes. The committee was asked to address three tasks related to Earth surface processes in the context of both scientific and societal issues:

1. Assess the current state and the fundamental research questions of the field of Earth surface processes;
2. Identify the rate-limiting challenges or opportunities for making significant advances in the field; and
3. Identify the necessary intellectual collaborations and high-priority needs to meet these challenges.

In this report, the field of Earth surface processes refers to the study of the form, physical properties, composition, function, and evolution of Earth's surface, a dynamic interface where physical, chemical, biological, and human processes cause and are affected by forcings in the Earth system, with impact-feedback loops that occur over a wide range of temporal and spatial scales. This report identifies nine grand scientific challenges that exemplify compelling directions of research in Earth surface processes (Chapter 2), and proposes four new, high-priority research initiatives designed to transform and strengthen the field in order to support the challenges (Chapter 3). The initiatives represent pathways to meet the demands for scientific information on issues related to planning, mitigation, and response to changes in Earth's surface now and in the future. Chapter 4 discusses the nature of the national support structure necessary to capitalize fully on these scientific opportunities.

The remainder of this chapter highlights some of the key advances and problems that have drawn attention to Earth surface processes research and contributed to its growth in the past several years. These examples focus on how Earth surface processes are interconnected or "coupled" to each other, to the atmosphere, and to the Earth's interior; on the increasing human impact on Earth's surface, including climate change; and on new technologies that have spurred recent theoretical advances in Earth surface processes. These topics are elaborated in greater detail in Chapters 2 and 3.

1.2 EXAMPLES OF INTERCONNECTED EARTH SURFACE PROCESSES

Climate, Tectonics, and Surface Processes

Interconnected processes at Earth's surface are coupled to those of Earth's interior in various ways that extend to millennial and longer time scales. The height and shape of rising mountains, for example, influence regional weather patterns, which affect rates of erosion via the amount and type of precipitation. As rivers and glaciers fed by topographically controlled precipitation carve deeply into uplifted rock in tectonically active areas, their concentrated erosion draws even more rock upward due to the effect of unloading (Figure 1.2). Spatial variation in erosion across a mountain belt due to climatic differences can affect the pattern of upward and lateral movement of rock toward Earth's surface. While the volume of rock drawn into a mountain belt is affected by Earth surface processes, the composition of the rock also is altered and this change can affect climate. Chemical weathering of rock freshly exposed in rapidly uplifting mountains affects the chemistry of water draining the mountains and can draw down carbon dioxide in the atmosphere, thereby influencing climate over relatively long periods of geologic time.

Even at these geologic time scales, biota are critical to the dynamic processes in mountain belts. Biotic processes mediate rates of rock breakdown (by weathering), soil development, and hillslope erosion and strongly influence the amount, size, and composition of sediment entering rivers. This sediment then influences the rate of bedrock incision, the geometry and dynamics of the channel, and the ecosystems that colonize an area.

Human-Landscape Dynamics

Largely within the last 3 millennia, humans have removed and replaced land cover, hastened the erosion of upland soils, and increased sediment supply to streams from upland erosion throughout many parts of the world (Figure 1.3). Worldwide damming of rivers has increased sediment trapping and residence times, however, greatly reducing the delivery of sediment to coasts and deltas. Although dams provide substantial societal benefits, including reduced flooding, hydroelectric power, and water for irrigation, their impact on sediment transport has caused the collapse of river ecosystems and starved coasts of sediment, leading to unanticipated delta subsidence, wetland loss, and greater coastal erosion.

Nearly every process on Earth's surface has been changed by human activities, heightening the need for new research on human-landscape dynamics and for a greater capacity to predict process responses to human influence. Earth-surface scientists have a unique and timely opportunity to use new tools and integrative approaches to enhance understanding and to predict future changes. More importantly, they are in position to transfer their

FIGURE 1.2 Linked tectonic, climatic, and Earth surface processes shape—and are influenced by—topography. Satellite image of New Zealand (*top*; February 9, 2002, SeaWiFS image) reveals regional climatic processes that drive spatial variation in rates and processes of erosion that, in turn, influence tectonics. Clouds are banked along the northwestern side of the mountains (Southern Alps) on the South Island; snow covers the northwestern peaks and higher slopes; glaciers occupy valleys draining high topography. Erosion, tectonic deformation, and uplift are focused on the western flank of the mountain range. SOURCE: Provided by the SeaWiFS Project, NASA/Goddard Space Flight Center, and ORBIMAGE. The schematic cross section of a convergent plate boundary (*bottom*) with similarities to the New Zealand example illustrates shortening and thickening of continental crust overlying the flowing mantle. The shape of the mountain range is influenced by climatically controlled, spatially varying erosion and its tectonic response (see also Section 2.3). Large black arrow indicates direction of motion of the converging plate. SOURCE: Modified after Willett (1999) and with permission of the American Geophysical Union.

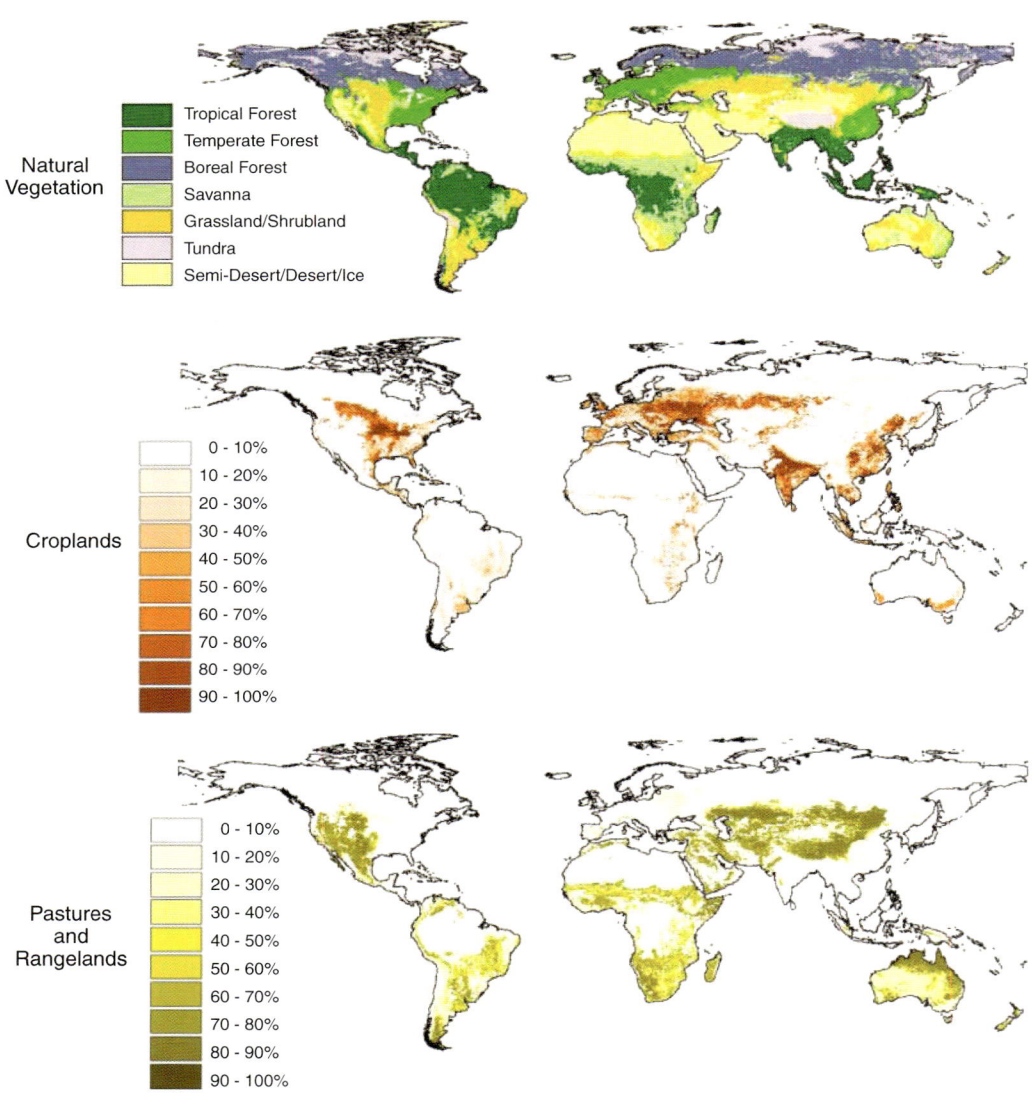

FIGURE 1.3 Humans have transformed nearly all of Earth's terrestrial surface. These maps illustrate the worldwide extent of human land-use and land-cover change: the geographic distribution of "potential vegetation" that would most likely exist in the absence of human land use (*top*); and the extent of agricultural land cover (including croplands and pastures) (*middle* and *bottom*) across the world during the 1990s. SOURCE: Foley et al. (2005). Reprinted with permission from AAAS.

knowledge to the greater scientific community, applied practitioners, the public, and policy makers in order to facilitate decision making.

As an example of the role of Earth surface science in providing greater understanding of Earth surface processes and in predicting systemic responses to change, consider what happens as aging dams are removed or breached. Tens of thousands of dams of various sizes have slowed the flow of rivers and trapped sediment and nutrients throughout the United States for up to hundreds of years, and dams continue to be built throughout the world. Removal of some of these aging impoundments is desired for reasons that include fish passage, human safety, and improved water quality. Yet pulses of sediment, nutrients, and pollutants are flushed from many breached reservoirs, impacting waterways, water quality, and habitat downstream. The nature and duration of downstream impacts are obvious questions of concern when dam removal is considered, but at present we are not able to predict accurately the changing rate at which sediment and nutrients will be eroded and transported downstream from a breached reservoir (Figure 1.4). What happens upstream of a breached dam is equally uncertain. Can the ecological, hydrological, and geomorphic functions of marshes, streams, and floodplains that lie buried beneath reservoir sediment be recovered after dam breaching? In essence, each dam breach is an experiment in which scientists can investigate interconnected Earth surface processes. Studying such experiments requires the expertise of scientists with diverse backgrounds and an ability to integrate analytical approaches, data, and interpretations across disciplines. The outcome of this kind of integrative research is invaluable to inform engineering practice and policy.

Soil Erosion

Natural soil erosion by the energy of wind, raindrops, or running water is accelerated by almost every human use of landscapes: agriculture, grazing, and timber harvesting, for example, may expose soil over large areas of continents to greater erosive forces and, in some cases, may degrade lands beyond beneficial use (Box 1.1). Soil eroded in this way may also accumulate in water bodies and create engineering or water quality issues associated with the transport of both nutrients and contaminants. Understanding and quantifying the magnitude of soil erosion and its downstream impacts across the United States are fundamental to inform policy decisions such as whether to subsidize soil conservation, to intensify cultivation for biofuel production, or to regulate land use in order to improve water quality in rivers, lakes, and estuaries. Nevertheless, considerable uncertainty exists about the magnitude of soil erosion and its downstream impacts across the United States. These uncertainties leave the nation in a fairly uninformed state that is exacerbated by inadequacies in the concepts presently used to make policy decisions related to soil conservation and land-use planning and illustrate the critical need for a better understanding of Earth surface processes.

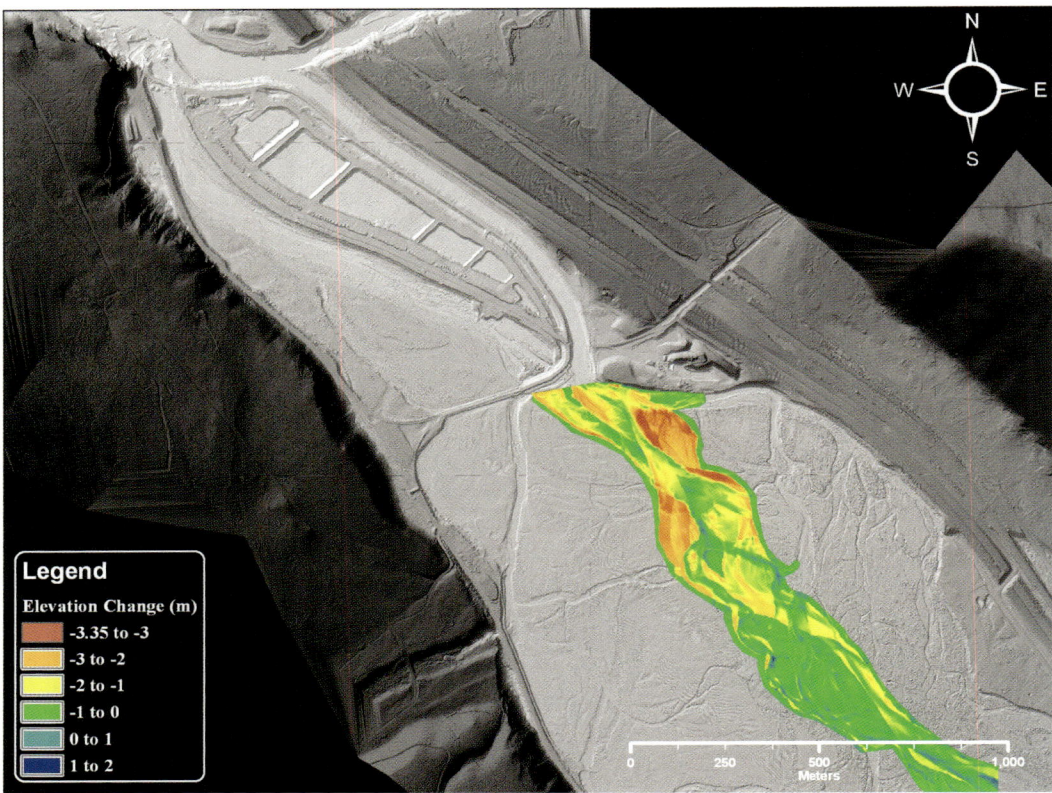

FIGURE 1.4 What happens when a dam is breached? Within months of dam removal (in upper left of image) in the spring of 2008, more than 180,000 m³ of sediment—far greater than predicted—was scoured from the upper part of the Milltown Reservoir, Clark Fork River, Montana. The Clark Fork River flows from lower right to upper left of image. Colors represent scour amounts determined by calculating the difference between digital elevation models of post-dam removal topography generated from an airborne lidar (light detection and ranging)[1] survey and of pre-dam removal topography from photogrammetry and bathymetric surveys. Immediately upstream of the dam site is a Superfund remediation area, where contaminated mining-derived sediments isolated from the river by berms are being mechanically excavated. Eroded sediment, which included arsenic from historic mining, was carried up to 200 kilometers downstream. SOURCE: Data from surveys commissioned by State of Montana; figure produced by and courtesy of Douglas Brinkerhoff and Andrew Wilcox, University of Montana.

---

[1] This report adopts the use of the lowercase form lidar for this acronym in keeping with the definition in Appendix C of the NRC report *Elevation Data for Floodplain Mapping* (NRC, 2007a). The lowercase word "lidar" is the appropriate form because it is directly analogous to "radar" and to a lesser extent "sonar."

Climate Change

A relatively recent human impact that has great import to Earth surface processes is climate change. Although climate change research has made significant advances in recent decades, examining the response of the Earth's surface to this change has just begun. Climate change affects all landscapes, influencing hydrology, flooding, water quality, nutrient loads, ecosystems, soil erosion, and landslide frequency. Retreating glaciers, for example, let loose large chunks of ice and freshwater while freshly eroded rock becomes exposed to weathering in the wake of the glaciers' retreating termini (Figure 1.5). An understanding of glacial mechanics, especially the basal sliding process, is important in predicting rates of glacial retreat and sea-level rise and is considered a top challenge for reducing uncertainty in climate projection and impact assessments.

Another important process in cold regions is thawing of the active layer of permafrost, which can produce nutrient and sediment pulses to coastal zones and increase the flux of carbon to the atmosphere. Permafrost underlies about 25 percent of global land area and is undergoing marked changes associated with recent global warming. Seasonally, the top of permafrost—the active layer—thaws and melts, and this seasonal thaw is increasing. The ecological impacts of warming and increased seasonal thawing of the active layer are complex, but potentially quite profound. Substantial amounts of carbon are stored in boreal soils, with possibly 50 percent or more of soil organic matter stored in high-latitude periglacial environments. Hydrologic conditions and biota play a role in whether thawed organic matter oxidizes or is reduced to methane. Although warming Arctic soils will likely be a source of carbon to the atmosphere, the details of carbon release from these soils are not yet clearly understood.

As with the need to address issues of glacial mechanics, understanding these fluxes in permafrost zones is critical to climate models that are sensitive to the changing concentration of greenhouse gases in the atmosphere. Coastal margins in the Arctic illustrate relatively short-term feedbacks between biota and deltaic channel-wetland systems that are linked to global warming and its effects on thawing of the active layer of permafrost. A vast delta (30,000 km^2) that formed at the mouth of the Lena River with sediment shed from northern Siberia currently contains extensive tundra wetlands that provide habitat for migratory birds (Figure 1.6). Biota and organic matter in the wetlands stabilize distributary channels and islands. Frozen for more than half of each year, these wetlands store large amounts of carbon that are released during thawing of the active layer in permafrost, which has been hastened by the warming climate. Organic-rich soils collapse as they warm and melt, leading to channel shifting, pulses of nutrient and sediment loading to the Laptev Sea, and the release of carbon dioxide and methane to the atmosphere. Such pulses can affect detrital food chains and benthic productivity along the continental shelf, which in turn affect the rate of erosion of coastal margins. The wetlands and frozen sediments of the

BOX 1.1
National Impacts of Soil Erosion and Sedimentation

Soil erosion, transport, and deposition are integral parts of the global sedimentary cycle and are important to altering and balancing the water- and nutrient-holding properties of soil profiles and transporting nutrients and sediment in and through rivers, reservoirs, estuaries, or deltas (see figure below). The needs for better methods of assessing and predicting soil erosion and sediment supply at a wide range of landscape scales, and for clarifying concepts that underpin public policy on these matters, suggest an important role for research in Earth surface processes. Major contributions from Earth surface research to assess the quantity and impacts of soil erosion involve computational models and a variety of measurement techniques applied to a range of erosive environments and timescales, and the ability to design and interpret these measurements through rigorous mathematical models of basin sediment budgets.

Much of the difficulty in interpreting and predicting the effects of soil erosion on water quality and sedimentation at whole-basin and national scales arises from several factors: (1) the difference between erosion at its source on hillslopes (as sheetwash, gully erosion, and landslides) and the full sediment budget of a basin; (2) the legacy of various natural and anthropogenic perturbations of river basin sediment budgets and the fact that a number of these perturbations have been quantified but largely ignored in assessments of soil erosion, transport, and sedimentation; (3) the general lack of rigor in measurements of soil erosion; and (4) a lack of realistic, quantitative estimates of soil production rates. The legacy perturbations in some regions include vast amounts of sediment that were mobilized by glaciers during the Last Glacial Maximum (~15,000 to 25,000 years ago) and redeposited in conditions and locations from which it is now being chronically remobilized by modern streams, waves of agricultural colonization, dam construction, and timber harvest. These changes in landscape evolution directly affect sediment budgets, and accurate assessments of soil erosion and sedimentation require incorporation of their effects. The actual measurement of soil erosion also requires greater rigor, particularly as it relates to the generalized statistical models generated from these data and subsequently used in land resource management.

A need exists to broaden the range of methods used to estimate soil erosion and to organize such measurements through landscape-scale designs and models that will allow realistic interpretation of the rate of soil erosion and the supply of sediment to streams and estuaries. Augmentation or rejuvenation of sediment sampling networks at river gauging stations is also important. The most important part of a strategy, however, would be to ensure that methods for measuring erosion, sediment transport, and accumulation are employed in a coordinated and representative manner with a view to quantifying basin sediment budgets.

With more accurate measurements, sediment budgets can then be quantified within the framework of mathematical models that allow the results to be checked for consistency and to be generalized for prediction purposes. Various basin sediment budget models with varying degrees of resolution and complexity are now being developed. A large-scale combined strategy of empiricism and model-based accounting will be required to answer these challenging public policy questions. The skills and tools are available in Earth surface research, which incorporates the whole range of scales of the problem from the production and mobilization of soils at a single point to the catchment- and continent-wide storage and transport of sediment.

The Importance of Earth Surface Processes

Human activities can lead to exceptionally high rates of landscape erosion. The challenge for modern studies of landscape change is to understand and measure the effects of these human activities relative to the background chemical and physical processes that also contribute to erosion and are driven by climate and tectonics. Madagascar is a case in point. Conventional wisdom holds that Madagascar's modern erosion rates are exceptionally high, and that the lavakas (gullies) which pepper the Malagasy landscape (pictured in the image) are the result of deforestation and local agricultural practices. New research, however, reveals this interpretation to be based on descriptive evaluations and regional extrapolations from few local datapoints. Recent studies, in contrast, show the importance of quantitative investigations at local to regional scales. Analysis of seismic events and mapping of gully distributions from satellite images, for example, show a very strong spatial correlation between earthquake distribution and the locations of dense lavaka clusters, underscoring the importance of tectonics in generating this complicated landscape. Ongoing measurement of sediment generation rates from modern and ancient river sediment will quantify changes in erosion rate due to anthropogenic disturbance. These broad-based studies are typical of modern approaches to understanding and quantifying the causes and effects of soil erosion and are central to the development of effective soil erosion control strategies. SOURCE: Cox et al. (2010); image courtesy of Rónadh Cox, http://drm.williams.edu/u?/lavaka,2141.

FIGURE 1.5 Global warming, retreating glaciers, and changing landscapes. The Columbia Glacier, flowing into Prince William Sound, Alaska, presently is undergoing rapid retreat that is likely to last another few decades. The glacier produces prodigious quantities of icebergs, which debouch into the sound near the shipping lane from the Valdez terminus of the Alyeska pipeline. The retreat is accompanied by thinning of the ice, revealing up to 400 meters of freshly exposed bedrock that bounds the present glacier. SOURCE: Photo courtesy of Robert S. Anderson.

Lena Delta system reveal the interconnectedness among climate change, biota, soils, and landscapes at the centennial to millennial scale and the critical linkages between human activities and Earth surface processes.

1.3 NEW TECHNOLOGIES: MONITORING EARTH SURFACE PROCESSES AT HIGH RESOLUTION IN SPACE AND TIME

The evolution and increasing availability of new measurement technologies has enabled many of the advances in Earth surface processes that are discussed throughout this report.

FIGURE 1.6 Interactions among climate change, biota, and landscapes. The vast Lena River delta, the largest delta in the Arctic region, formed as sediment from the Lena River was deposited where it flows into the Laptev Sea. Tundra wetlands in this delta store large amounts of carbon that potentially could be released by modern global warming. Attributed largely to human activity, warming accelerates permafrost thawing and the erosion of organic-rich delta sediments. Envisat image acquired on June 15, 2006; width of image ~350 kilometers. SOURCE: European Space Agency.

Technological advances in remote sensing, geochemistry, geochronology, and computing have fostered great progress in the study of Earth's surface (Appendix C). For example, recent advances in the areas of digital topography and geochronology enable scientists not just to conduct research faster or more accurately, but to make observations and interpretations that were not possible previously.

Digital Topography

Throughout history, the creation of maps has been a means of recording observations that enable us to find and denote paths and patterns and to generate hypotheses about the controls on the spatial relationship of features. Topographic maps, depicting land elevation and displaying landforms, have been crucial to scientific inquiry about the Earth and have been central to land development. In the 1980s, a profound step was taken when line drawings of elevations on topographic maps were digitized and the landscape could be represented via digital elevation models on computers. This innovation launched thousands of scientific studies exploiting this new capability and ultimately gave rise to many new practical applications. In the last decade, technological advances have enabled the first airborne and satellite-mounted surveys of topography using radar (interferometric synthetic aperture radar, InSAR) and laser (light detection and ranging, or lidar) technology, giving unprecedented spatial resolution over large areas. This development has led to a second wave of digital topographic studies that are transforming not just research in Earth surface processes, but also the fields of agriculture, ecology, engineering, and planning. With regard to Earth surface processes, digital elevation data enable us to examine, for the first time, topographic features over broad areas using computer-automated techniques. This ability is leading to new insights and tools that link landscapes to hydrology, geochemistry, tectonics, and climate. Although many digital elevation data are coarse in scale for studying the features, for example, of mountain belts with long, high hillslopes, the data have been truly revolutionary. The advances in the past decade are akin to those of the 1960s in the fields of seismology and geophysics, when accessibility to global seismic and paleomagnetic data and new tools to process such signals spurred the plate tectonics revolution and greater understanding of Earth's subsurface processes.

One of the most recent transformative phases in the measurement and characterization of landscape topography has been the ongoing development of laser surveying, both from the ground and from airborne instruments. This method is referred to as lidar, or airborne laser swath mapping (ALSM) in the case of aerial surveys. High-resolution swath bathymetry uses sonar for the same types of measurements in marine environments. With lidar, a laser pulse is sent from the instrument, and the time for its return from a reflected surface is detected and used to calculate distance. Current technology permits typical accuracies to about 5 to 10 centimeters vertically and 20 to 30 centimeters horizontally, with data returns every few decimeters. From these returns a point cloud of elevation data is created; various analytical methods are then used to distinguish vegetation from ground (Figure 1.7).

For the first time, we now can obtain surveys over broad areas that document topography at the resolution at which transport, erosion, and deposition processes operate. Lidar data also capture important quantitative attributes of vegetation that can be used in studies

The Importance of Earth Surface Processes

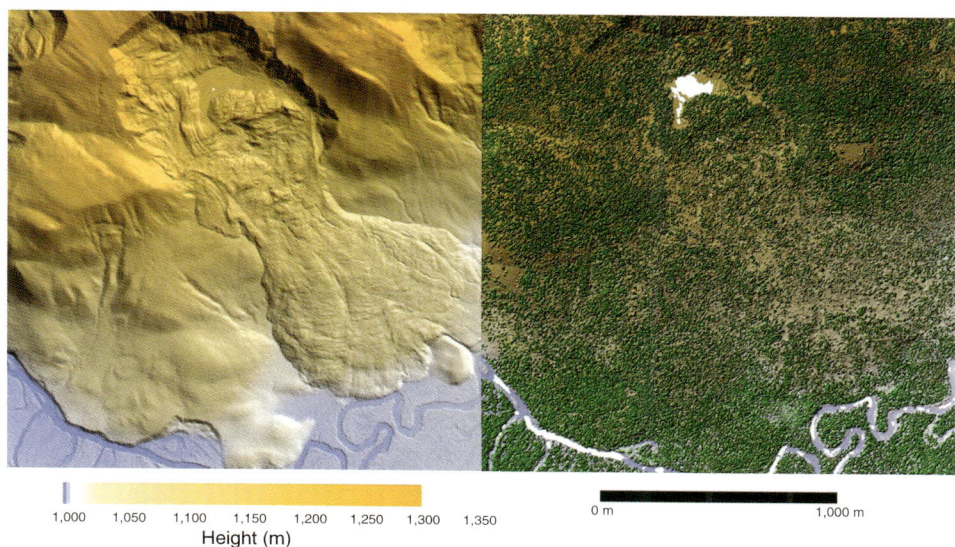

FIGURE 1.7 Documenting topography at the resolution of transport and erosion processes. Comparison of Google Earth image (*bottom*, digital air photo) with two lidar-derived images for an area near Flathead Lake, Montana: "bare Earth" topography with vegetation removed (*top left*) and vegetation color-coded by height above bare Earth (*top right*). Area in red box on the digital air photo is the area covered by the upper lidar images. Image width of the digital air photo is ~3 kilometers. SOURCE: National Center for Airborne Laser Mapping (NCALM), Bottom Image courtesy of Google Earth.

of ecohydrology and ecogeomorphology. Landslide scars, channel banks, river terraces, floodplain features, fault traces, and other landforms can be detected, quantified, and used to advance theoretical and practical understanding. Repeat scans allow change detection as never previously possible. These techniques also permit improved understanding of the human impact on types and rates of geomorphic processes.

Geochronology

To quantify rates of Earth surface processes and ages of landforms, Earth scientists have developed in the past 20 years a wide range of tools that exploit the time-dependent exposure of materials to cosmic rays, heat, and light. The greatest breakthrough came when measurement techniques advanced to the stage that trace concentrations of atoms produced by cosmic rays could be isolated and measured accurately (Box 1.2). Questions posed by early twentieth century Earth scientists about ages of landscapes and their evolutionary sequences, and the underlying mechanisms of erosion and deposition, can now be addressed quantitatively. These rate measurements coupled with new thermochronometers (see Box 2.4) have revealed suspected but previously unmeasurable linkages between erosion and tectonics. Undoubtedly much more will be discovered as these new dating technologies are used to measure the rates of evolution of Earth's surface.

1.4 STUDY CONSIDERATIONS AND REPORT STRUCTURE

Interdisciplinary research[2] in Earth surface processes comprises the detailed investigation of contemporary processes that generate and degrade landscapes and change the properties of rocks and soil; the definition of how these processes have functioned over the long periods of time required for the evolution of surface conditions (composition, function, and form); the deep connections among surface processes, climate, tectonics, life, and human activity; and ultimately the prediction of future landscapes and the fluid, solid, and solute fluxes across them. Evidence of the environmental history of landscape development is stored in the geologic and geochemical records of sediments, water, and soils. Building from datasets that extend across space and time and using a growing variety of powerful tools and techniques, scientists are able to measure landforms, probe sediments and

---

[2]This report has adopted the definition of "interdisciplinary research" as provided in the National Research Council (2004) report *Facilitating Interdisciplinary Research*. The report defines interdisciplinary research as a "mode of research by teams or individuals that integrates information, data, techniques, tools, perspectives, concepts, and/or theories from two or more disciplines or bodies of specialized knowledge to advance fundamental understanding or to solve problems whose solutions are beyond the scope of a single discipline or area of research practice." The report goes further to indicate that the interactions between fields or disciplines may result in forging a new field or discipline.

water, quantify process rates, and model the changing face of Earth. As interdisciplinary approaches increase the power of landscape research, a complex picture is beginning to emerge of landscape functioning, evolution, and interactions with life and human activity. Research in this area is integrative because it involves linkages to many related fields and because the core of the research lies at discovering the interactions and feedbacks involving physical, chemical, biological, and human processes.

The field of Earth surface processes overlaps with studies of the Critical Zone as defined by the NRC report to NSF on *Basic Research Opportunities in Earth Science* (BROES) (NRC, 2001a). The BROES report first developed the concept of the "Critical Zone" as the "heterogeneous, near-surface environment in which complex interactions involving rock, soil, water, air, and living organisms regulate the natural habitat and determine the availability of life-sustaining resources." In addition to these surface interactions, the investigation of Earth surface processes as developed in this report involves features and transfer processes that place greater emphasis on geological history and on interactions and feedbacks both with humans and with deep-Earth processes (e.g., tectonics) than those initially conceived in the definition of Critical Zone studies. Notably, the NSF-supported Critical Zone Observatories, which have grown out of this interest in the Critical Zone, are an integral part of the effort to advance the understanding of processes operating at the surface of the Earth (see Box 2.5). In this report, the committee did not develop static definitions of Critical Zone science or Earth surface processes, but considered them part of the same urgent effort to understand processes operating at or near Earth's surface.

To identify most effectively the greatest challenges and most promising opportunities in Earth surface processes, the committee sought input in its public meetings from panelists whose expertise augmented that of committee members (Appendixes A and B) and remained abreast of the meeting activities of a parallel, but separate, NRC study *Strategic Directions in the Geographical Sciences*. The committee also sought input from a broad section of the international scientific community relevant to Earth surface processes through an online questionnaire (see Appendix B). Responses to the questionnaire emphasized a number of recurring themes including (1) the interconnectedness of diverse processes acting on Earth's surface; (2) the importance of incorporating human dynamics in research on Earth surface processes; (3) the value of new technologies to advance understanding of Earth surface processes (see also Appendix C); and (4) the scientific and practical challenges that face the community in its effort to advance this field. Although this report focuses on the terrestrial surface, the committee emphasizes that research on the submarine surface and on marine processes is as active and exciting as terrestrial research. However, without any marine scientists on the committee, we did not have the resources to do justice to topics addressed by this important, allied community.

Numerous boxes and figures throughout the report are designed to highlight specific concepts, tools, and examples related to Earth surface processes research that otherwise are

BOX 1.2
The Cosmogenic Radionuclide Revolution

Earth is bombarded constantly by high-energy cosmic rays such as protons and neutrons. Cosmogenic isotopes are rare isotopes that form in Earth's atmosphere and near surface when cosmic rays hit the nucleus of target atoms such as ^{16}O, ^{14}N, ^{40}Ca, and ^{28}Si. The target atoms are abundant in rock, soil, and the atmosphere. The rates of production and decay of many cosmogenic nuclides are known sufficiently well that measured nuclide concentrations in samples can, among other things, be used as an absolute dating technique. The production of cosmogenic nuclides in a wide range of Earth materials and knowledge of the production rates have led to a revolution in our ability to quantify the timing and rates of Earth surface processes from thousands to millions of years with radiogenic nuclides (^{7}Be, ^{10}Be, ^{14}C, ^{26}Al, ^{36}Cl), or even longer with stable nuclides (^{3}He, ^{21}Ne), in a diverse range of environments and over a wide range of time scales (see figure below).

Most Earth surface applications of cosmogenic nuclides include either dating the exposure time (or age) of a surface to cosmic rays and/or quantifying landscape denudation rates. Examples of exposure age dating include dating glacial moraines to quantify the timing and magnitude of ice ages, dating of groundwater to understand long-term migration velocities and the effects of climate changes on recharge, and dating of fluvial terraces, alluvial fan surfaces, and landslide deposits to investigate the effects of tectonic processes on crustal deformation, faulting, and erosion processes (see figure below). Denudation rates or magnitudes can be quantified by measuring nuclide concentration in modern or ancient fluvial sediments for catchment average rates or from bedrock samples at points across a landscape. The ages of sediments can now be documented to within a few years over the last two centuries, and sediments from individual floods can be identified during the following season. More recently, cosmogenic nuclide concentrations in soils have been used to understand the rates and magnitudes of soil mixing due to biologic and periglacial processes and soil production rates from bedrock.

Accelerator mass spectrometry (AMS) is required for measurement of most radionuclides. NSF, and other science programs globally, have committed funds to the development and maintenance of AMS facilities. Subsidized facilities such as the Purdue Rare Isotope Measurement (PRIME) laboratory at Purdue University keep the costs of data collection reasonable for the benefit of all.

The future of cosmogenic isotope geochemistry is promising. Examples of frontiers in this field include (1) refinement of nuclide production rates through the United States and European Cosmic-Ray produced NUclide Systematics on Earth (CRONUS) and CRONUS-EU initiatives to reduce uncertainties in exposure age and denudation rate calculations, (2) development of significantly smaller and more affordable AMS facilities to reduce costs and increase sample throughput, and (3) continued development and application of cosmogenic noble gas techniques to a wider range of minerals. Of particular interest is recent progress in measuring cosmogenic ^{3}He in common mineral phases that are present in many rocks and are more retentive of helium than quartz.

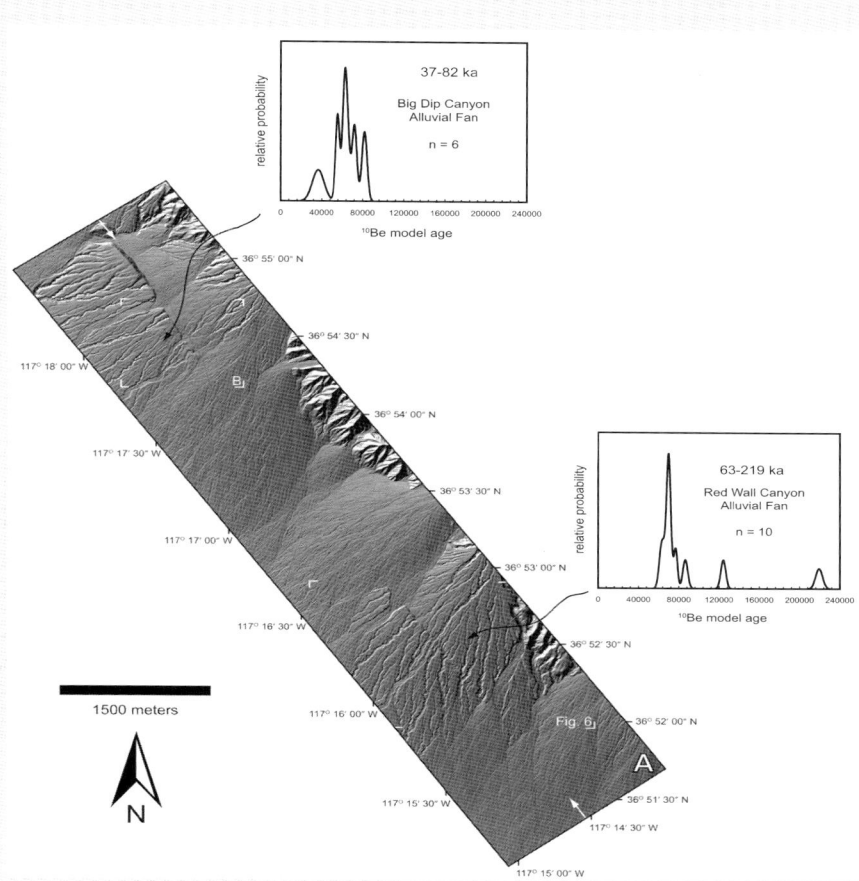

Dating landforms and determining fault slip rates. Different ages of alluvial fan deposits form surfaces with varying degrees of surface roughness, desert varnish, and other indicators of relative age. Cosmogenic dating, in this case ^{10}Be, provides chronologic control for different ages of alluvial fan surfaces offset along the northern Death Valley fault (fault strikes northwest to southeast along the western part of a lidar-derived bare-Earth topographic image). The development of these fans has been linked to changes in climate. The age estimates of these features are used to reconstruct the offset fan surfaces along the fault in order to calculate a slip rate of 4.5 millimeters per year. SOURCE: Frankel et al. (2007) and with permission of the American Geophysical Union.

mentioned only briefly in the text. Neither these boxes nor the examples in the body of the text can represent the entire range of research covered by the field, but they are intended to draw attention to contributions from various disciplines to further research on Earth surface processes. Similarly, we do not cite many specific research publications from among the vast number in this area of research except to provide proper attribution for a point of fact, a figure, a direct quotation, or an explicit concept. The committee also notes that the study charge is focused on research and, for this reason, has deliberately not included education and human resource issues in its discussion. Although clearly of importance to all areas of science and engineering, capturing education and workforce topics in adequate detail was beyond the study scope. Appendix D provides some background to the growth in this field at universities and in the international professional community.

In addition to the collaborative and integrative approaches emphasized in this report, the committee recognizes that the emerging science of Earth surface processes often has relied on fairly simple, descriptive approaches. Empirical methods have underlain much of the theory development for understanding landscapes, and observations and data collection will remain important components of studies of the Earth's surface. Nevertheless, significant advances will require developing quantitative predictive capability for how landscapes form, evolve, and respond to change. Such capability is especially important as Earth surface processes are increasingly altered by human activity and climate change. For the foreseeable future and for most landscape processes, predictive models will of necessity continue to be partially empirical, even as improvements in understanding underlying processes gradually are achieved. A useful analogy here is one of weather forecasting, which combines sophisticated numerical solutions of the governing equations of atmospheric dynamics with empirical relations for incompletely understood processes (e.g., cloud formation) to make forecasts that are quantitative and, at the same time, stochastic. Although the underlying equations for landscape evolution are not known at present and may be quite diverse, the general approach to landscape prediction is likely to be similar.

1.5 CLOSING REMARKS

Intended for use by NSF, decision makers, and research communities in academia, the private sector, and federal agencies, this report identifies high-priority areas for research in Earth surface processes. Many of the research areas address critical societal needs and issues. The report also suggests means to coordinate and support the necessary research. Basic research in Earth surface processes is one part of the research portfolio encompassed by the Section of Surface Earth Processes at NSF. Many of the exciting research activities and intellectual advances in the study of Earth surface processes, however, have gone beyond traditional disciplinary boundaries of Earth science, reaching into research domains that fall within other areas at NSF, such as climate, ecosystem sciences, tectonic processes, and

Earth's interior. The field of Earth surface processes lies at the intersection and integration of diverse natural science disciplines—Earth, life, atmospheric, and ocean sciences—that address explicitly the function, composition, form, and evolution of Earth's surface and near-surface environment. Increasingly, as the human impact strengthens, the field of Earth surface processes requires integration with the social and behavioral sciences as well.

Earth's surface is the only habitat available to the human race. Understanding the processes by which that habitat has been created and continually altered is important to determine the causes of environmental degradation, to restore what is degraded, and to guide policy decisions toward a sustainable Earth surface. The agencies and individuals responsible for natural resources and public welfare widely acknowledge that environmental change occurs on a scale and with an intensity that is important for society's long-term plans and investments. This acknowledgment now puts a responsibility on the shoulders of natural scientists as well as professionals in economics, social science, engineering, and other fields to interpret the record of ongoing environmental change and to anticipate and, in some cases, make quantitative predictions of future events or conditions. Earth-surface scientists who study such records have distinctive insights, methods, and skills to understand the form, composition, properties, function, and evolution of Earth's surface and to contribute to resolving modern environmental challenges.

We have the technological ability to monitor closely human impacts on the environment. The need to observe, measure, and model human-landscape interactions in an integrated, predictive fashion is clear. To develop this capability, fundamental research is needed to understand and quantify the impact and feedback relationships between human activity and Earth surface processes. In many cases, the socioeconomic will and capacity exist to attempt to alter the impacts and responses initiated in human-influenced landscape systems. With new scientific questions about various components of the Earth system, opportunities and tools for research, rapid growth of the human population, and unprecedented changes in biota, land cover, process rates, and global climate, an appraisal of the study of Earth surface processes is both timely and crucial.

CHAPTER TWO

Grand Challenges in Earth Surface Processes

A number of overarching intellectual questions regarding natural and anthropogenic processes are fundamental to understanding the form, composition, and evolution of our planet's surface. From among these questions, the committee has identified nine grand research challenges, each of which is introduced in the form of a primary research question with associated opportunities for further investigation. The grand challenges are exemplary, but they cannot provide a full assessment of all current research in Earth surface processes. Focus was given, instead, to integrating a range of topics that are representative of the breadth and depth of research in the field. The nine challenges are supported by information gathered from the community of Earth surface scientists[1] during this study (Appendix B).

2.1 WHAT DOES OUR PLANET'S PAST TELL US ABOUT ITS FUTURE?

One of the most remarkable aspects of the Earth surface system is the extent to which it records its own evolution. The visible landscape incorporates a wealth of information about its past, and the archive of sediments and sedimentary rock extends the record to very early in our planet's history (Figure 2.1). The Earth's record keeping has given us, among other things, a detailed picture of the evolution of the atmosphere, oceans, tectonics, life, and the surface system itself. Both scientific curiosity and societal demand drive us to read this fragmentary and often puzzling archive in an attempt to understand how the surface environment in which we exist has evolved through time and therefore how it may change

---

[1] As used in this report, the term "Earth surface scientists" refers to scientists from disciplines concerned with the form, composition, properties, function, and evolution of Earth's surface, including biogeochemistry, ecology, environmental science, geochemistry, geography, geology, geomorphology, glaciology, hydrology, oceanography, sedimentology, soil science, and stratigraphy. Increasingly, scientists from disciplines such as atmospheric science, engineering, geophysics, and social science are participating in research on Earth surface processes as well.

FIGURE 2.1 The stratigraphic section in the Grand Canyon records a time span from about 1.8 billion to 250 million years ago—a substantial fraction of Earth's history. SOURCE: National Park Service.

in the future. Our growing ability to quantify processes specific to the near-surface environment also lets us reconstruct Earth's past in increasing detail including aspects of its geochemistry, biotic processes, topography, and particle and solute fluxes. The information stored in landscapes and sediments provides a kind of "time telescope" that allows us to view what is in effect a temporal sequence of alternative Earths—recognizably our planet, yet often surprising and unfamiliar.

In addition, the history of landscapes strongly influences their present state and future evolution. In the Earth's northerly regions, the current landscape pattern was largely created by glacial processes that peaked around 18,000 years ago, and present-day debates about agricultural best practices and practical water quality standards require a sophisticated understanding of how the landscape is evolving in response to the deglaciation. Likewise, coastal erosion is still influenced by uplift and subsidence in response to shifts in ice loading over Holocene time. The fate of global deltas on which hundreds of millions of people depend in turn depends on the delicate interplay of subsidence and sedimentation developed over geologic time. In upland regions, the general importance of tectonic history is obvious; more subtle are the possible effects of variations in uplift rate and climate in influencing the balance of sediment storage and release to downstream river systems. Soils are among the clearest examples of the influence of past time—the soils that support global agriculture represent the integrated effects of tens of thousands of years of biogeochemical processes.

The evolutionary road from the prebiotic Earth and an atmosphere devoid of oxygen to the human-dominated conditions of today has been well documented in the popular press: for example, extreme variations in global ice cover from nearly ice-free to (some believe) near-complete cover; meteorite impacts; sea-level variations of hundreds of meters; supercontinent assembly and breakup; mass extinctions; the growth and decay of mountain ranges; and the rich variety of life forms that preceded human existence. The sedimentary record of Earth's colorful past is a composite of self-recording landscapes. Beyond the drama and fascination of Earth history, the record in landscapes and sediments gives us an archive of natural experiments, performed at full planetary length and time scales, from which to reconstruct planetary dynamics. This deep-time record of landscape evolution also opens an avenue to explore the connections between the Earth's interior and its surface boundary. The morphology and properties of Earth's surface constitute fundamental, directly observable attributes that help constrain the long-term behavior of the Earth's convecting interior and the tectonic plates that gradually move across its surface. Landscapes on other worlds also provide a chance to test and expand our understanding of the Earth's surface processes by applying them to materials and basic surface conditions (e.g., gravity, atmospheric pressure) very different from the ones with which we are familiar.

Practical reasons also exist for studying the record of surface evolution. Nearly every aspect of the present form, composition, and function of Earth's surface reflects its evolution over geologic time; thus, the surface environment—including the services it provides and the hazards it poses—cannot be understood outside the context of its history. Second, the subsurface heterogeneity that controls the availability of resources such as water, hydrocarbons, and minerals is in effect a three-dimensional tapestry of fossilized surface dynamics created via crustal subsidence and burial of paleo-landscapes in sedimentary basins. Lastly, the history of surface evolution in landscapes and sedimentary basins provides us with a rich archive from which to extract information on extreme events, natural variability in space and time, and how the surface responds to change. This archive is sometimes the only means we have to observe critical thresholds and other forms of nonlinear response to our modifications of the Planet's surface systems, such as those induced by human activity.

The past as the context for the current environment. Although "sustainability" has been defined in different ways, at its core it connotes management with a long, perhaps indefinite, time horizon. The further ahead we would like to forecast, the further back must we look to understand how the system arrived at its current state and how it is likely to evolve. For example, consider two contrasting scenarios for beach retreat, an issue that has consumed billions of dollars in the United States. One is a case of local erosion caused by interruption of longshore sediment flow by jetties. In the other, the retreat is the result of long-term land sinking caused by adjustment from the last glaciation. Although the short-term responses to both cases might appear similar, their contexts and hence strategies for sustainable management may be entirely different.

Scenarios such as this, in which long-term processes strongly influence the modern environment, can be found across the Earth's surface. Centennial-scale or longer variations in seismicity and climate may control landslide frequency in steeplands, altering the supply of sediment to river systems, with a broad spectrum of environmental consequences. Over much of the Northern Hemisphere the strong imprint of glacial history persists in the distribution of soils, surface biota, river courses, and groundwater. The extraordinary fertility of the breadbasket of the United States is related to soils developed on sediments deposited in the wake of glacial retreat; now human activities are changing the physical structure and geochemistry of these soils and locally stripping them rapidly from the landscape (Box 1.1; see also Sections 2.4, 2.8, and 2.9).

As long as we thought we could simply control the landscape for the convenience of humankind, temporal trends such as these were primarily of academic interest. Now momentum is growing around the world to move away from this type of "hard" engineering to a more sustainable approach based on working with natural tendencies rather than against them. This cannot be done unless we know what the natural tendencies are—a task that requires a new synthesis of process understanding with research to reconstruct past history quantitatively.

Subsurface architecture. Buried surface features such as channels and beaches serve as conduits and reservoirs for water as well as for oil and natural gas (Figure 2.2). In addition, the new field of carbon sequestration focuses on attempts to remove carbon dioxide (CO_2) from the atmosphere and store it indefinitely in subsurface reservoirs—often ones from which hydrocarbons have been extracted. Again, this cannot be done safely and sustainably without detailed knowledge of the subsurface plumbing system.

The petroleum industry spends billions of dollars creating extremely detailed images of the subsurface architecture created as basin subsidence and sediment input build depositional sequences in three dimensions (Figure 2.3). These images are used to characterize hydrocarbon reservoirs, but as records of geomorphic evolution, their potential value for understanding surface dynamics is enormous and has hardly been tapped.

Four billion years of natural experiments. Given the complexity of the surface environment, with its interwoven and highly nonlinear physical, geochemical, and biotic systems, it is not surprising that theoretical methods for predicting its evolution are only in their early stages. In addition to hypotheses, we must learn how the Earth's surface works through field observation and experimentation. Intentional scientific experiments in the field are common, but necessarily involve short length and time scales; reduced-scale laboratory experiments are proving to be useful, but they cannot capture all aspects of surface dynamics. Humankind seems determined to carry out full-scale unplanned "experiments," through activities such as changing land use, reconfiguring and regimenting natural landforms, and

FIGURE 2.2 Buried channels such as these in the Ebro Basin, Spain, continue to serve as preferred conduits and reservoirs for fluid flow (oil, gas, and/or water) in the subsurface. SOURCE: Photo courtesy of Christopher Paola, University of Minnesota.

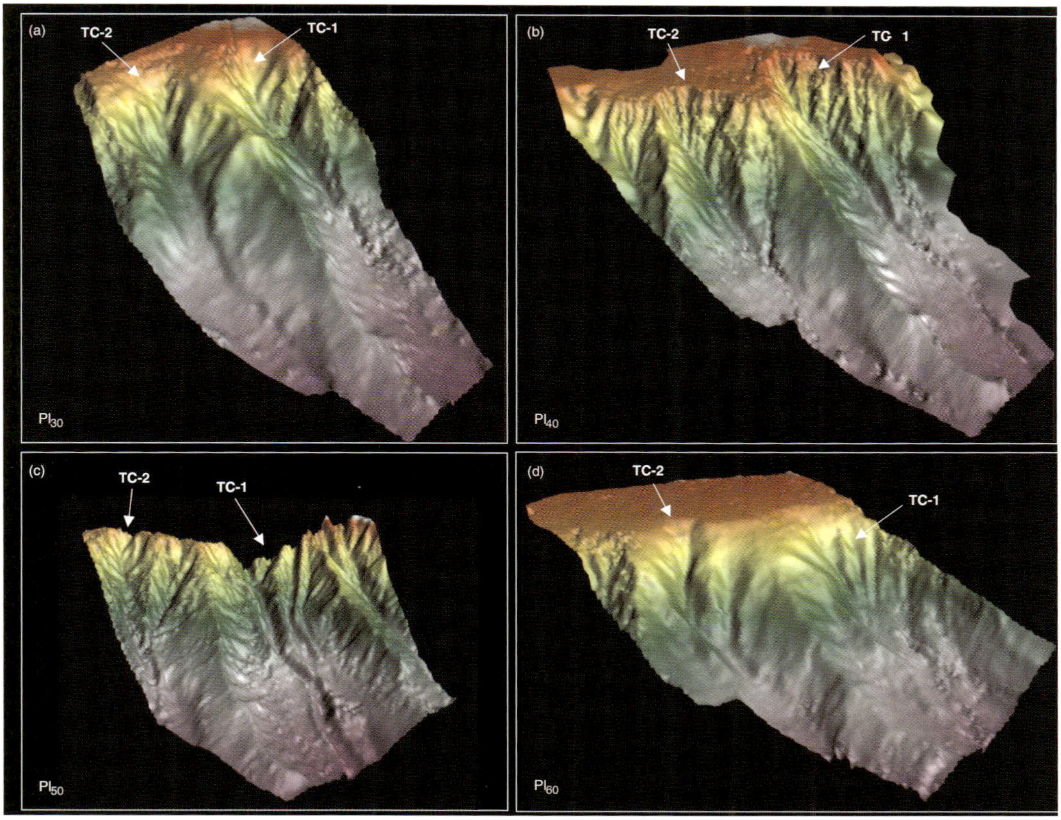

FIGURE 2.3 High-resolution seismic reflection has opened the way to "seismic geomorphology," including temporal evolution of features like the submarine landscapes visualized here. These topographic maps (field of view 8 × 12 kilometers, relief of approximately 600 meters, with 4 times vertical exaggeration) represent a time series of Pliocene (circa 3.5 million years ago) seascapes (the oldest is A, the youngest is D) now buried at depth offshore of the Ebro Delta in the northwest Mediterranean Sea. Reflections from deeply buried surfaces were extracted from a three-dimensional seismic volume to produce these maps of ancient seascapes. SOURCE: Bertoni and Cartwright (2005). Reproduced with permission of Blackwell Publishing Ltd.

altering the climate. We will learn from these experimental results even as we try to predict and perhaps influence them. Yet the increasing influence of humans at global scales impels us to make better use of the roughly 4-billion-year record of natural experiments that have already occurred on Earth. The record of natural experiments includes extreme events such as meteor impacts and rapid climate changes, as well as states of the Earth—for example, one that is nearly ice-free—that are quite different from the one we know. Although we have much to learn, we do know that the world of direct human experience represents only a small fraction of our planet's highly varied history and of its potential future. Now as we put increasing pressure on environmental services, urbanize hazard-prone landscapes, and become predominant geologic agents ourselves, understanding the full range of planetary behavior has become crucial. Numerous other examples of insights gained from studies of surface evolution on a range of time scales are included in other sections of this report.

The archive of Earth history records a far richer range of states and behaviors of the landscape than we can observe directly in the tiny slice of time we occupy today. As such, sedimentary and other landscape records can provide critical information about how the landscape could change in the future. Three primary research opportunities to understand Earth's surface and its future through examination of its past include (1) quantitative reconstruction of landscape records, (2) application of dating and imaging tools to understand landscape history, and (3) linking studies of the surface and subsurface.

Quantitative Reconstruction

Interpretation of landscape records by Earth surface scientists to this point has been mostly qualitative. The crucial next step is to accelerate the process of learning to read these often fragmentary records quantitatively, in order to understand long-term dynamics and reconstruct extreme events and variability and to provide the results in a form that can be used by decision makers. Recent examples of reconstruction include the spectacular findings from ice-core analysis showing that climate can change dramatically (see also Section 2.7)—over time scales measured in years—and from quantitative estimates of storm and earthquake frequency from sediment studies. These inspiring results represent a first step to unlocking the surface and sedimentary archive. One way to focus this type of reconstruction work is to seek "distant mirrors"[2]: times in Earth's past with high potential to shed light on problems of prediction facing us today, such as accelerated climate change. Initiatives to study the Paleocene-Eocene Thermal Maximum (approximately 55 million years ago), an ancient warm interval potentially analogous to an anthropogenically warmed Earth, provide a good example of this approach.

---

[2] Quoted material credited to B.W. Tuchman, 1978. *A Distant Mirror: The Calamitous 14th Century*. New York: Random House.

Landscape History

New imaging methods allow us to study the present morphology of the landscape in unprecedented detail (Box 2.1; see also Sections 2.2, 2.3, and 2.6; Appendix C). Equally importantly, new radiometric and other surface-dating methods (Box 1.2; see Sections 2.2 and 2.3; Appendix C) allow us to measure rates of surface change and add temporal evolution to our snapshot of the current state of the landscape. These tools are precisely those needed to quantify landscape evolution and to reconstruct the path that has led to the present state. In tectonically active areas, analysis that includes data from these tools can, for instance, extend seismic records to time spans beyond those of human records (see also Section 2.3). In any area, reading geomorphic history quantitatively will allow us to measure fluctuations in key environmental quantities, such as fluxes of sediment and nutrients through Holocene time (the last 11,700 years of Earth history), and to constrain the magnitudes of extreme events (see also Section 2.7). These essential data provide the natural context for the observed acceleration of anthropogenic changes to the surface environment.

BOX 2.1
Digital Landscapes: Documenting Topography at the Scale of Transport and Erosion Processes

Airborne laser mapping began only in the mid 1990s, and the technology has improved rapidly since then. In 2003 the National Science Foundation (NSF) supported the founding of the National Center for Airborne Laser Mapping (NCALM) to provide research-grade airborne laser swath data to the national research community, to advance the technology, and to provide education and training for students. By January 2009, more than 60 projects covering some 13,000 km^2 had been flown. For the GeoEarthScope program, all of the San Andreas Fault System (~3,000 km^2) was surveyed, and those data (plus other fault zones flown by NCALM) are freely available via the GEON OpenTopography Portal (http://www.geongrid.org/index.php/topography/opentopo). NCALM makes available the data for all its surveys through its web page (http://calm.geo.berkeley.edu/ncalm/ddc.html). By the end of its first 10 years, NCALM will have flown about 80 seed projects in support of graduate research, providing key data and contributing to a new generation of researchers advancing the field of Earth surface processes through use of high-resolution topography. In 2008, NSF supported an international workshop on High-Resolution Topographic Data and Earth Surface Processes to explore how high-resolution topographic data can advance understanding of Earth surface processes (Merritts et al., 2009). Nearly half of the more than 60 participants were graduate students or postdoctoral researchers who had received a doctorate within the past three years.

Connecting Surface and Subsurface

For too long the closely related sciences of geomorphology and sedimentary geology have been pursued separately. The confluence of the need to understand and predict the evolution of the surface environment and the availability of large volumes of high-resolution information on the subsurface archive—much of it from the oil and gas industry—provides a new level of motivation to bring these two fields together both to improve prediction in the subsurface and to use the archive to understand how the surface system works. Concerted effort to understand the links between the Earth's surface and its deep interior also provide an opportunity to meet this challenge (see also Section 2.3). Another key element is to develop observations and methods that link processes across the range of time scales from the present day to deep time. Laboratory experiments that, in effect, speed up time are one way of approaching this. Another bridge to deep time is the set of natural climate experiments constituting the current "ice house" (glacially influenced) period. These experiments are being investigated intensively for their value in understanding climate dynamics and testing numerical climate models.

2.2 HOW DO GEOPATTERNS ON EARTH'S SURFACE ARISE AND WHAT DO THEY TELL US ABOUT PROCESSES?

A glance out an airplane window on a clear day is enough to remind us of the remarkable capacity of landscape processes to create spatial patterns. These *geopatterns* comprise a diversity of scales and forms, and most show a fascinating mix of order and disorder. Familiar examples include the treelike, branching patterns of stream networks that create erosional and depositional landscapes; river channels, with their ornate meanders and braids; sand dunes; glacial valleys and landforms; deltas; barrier islands; and the zones and fabrics of soils (Figures 2.4-2.6). Physical landscape patterns are often closely associated with biotic ones, ranging from the variation in forest type with upland elevation to riparian ecosystems tied to stream channels to the exquisite control of marsh vegetation by small changes in land elevation and wetting frequency. In inhabited landscapes, we notice the human footprint, visible as clearly "unnatural" patterns (Figure 2.7; see also Section 2.8). These include cultivated areas with simple geometric boundaries quite unlike the intricate patterns of natural landscapes, as well as cities and towns that may exhibit locally regular spatial structures.

New kinds of surface geopatterns appear as we explore landscapes that are new or are on unfamiliar scales. For example, advances in sonar and other underwater imaging techniques have enabled us to visualize underwater landscapes as if from an airplane, revealing spatial patterns on the seafloor that often appear to be scaled-up cousins to their terrestrial counterparts (Figure 2.8).

Grand Challenges in Earth Surface Processes

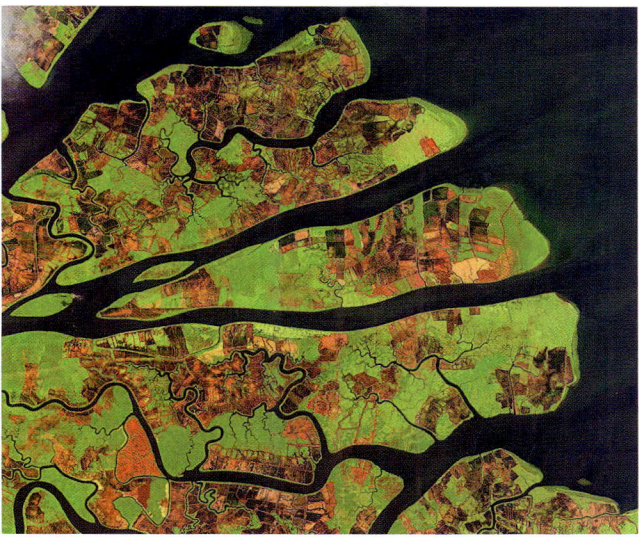

FIGURE 2.4 These two images show that repeating patterns apply in different landscape types. These and many other geopatterns arise through local interactions and structure the landscape. What do they tell us? The image on the left shows patterns in a mountain and valley landscape on the border of China and Myanmar, while the image on the right shows the sinuous and branching patterns present at the mouth of the Kayan River, Indonesia. SOURCE: SPOT satellite image © CNES (2009), acquired by CRISP, NUS.

Microscopic imaging has also revealed patterns of mineral dissolution at micron scales that are similar to those developed on the scale of landscapes. Measurement of molecular biological signatures has revealed the details of spatial patterns, recognized for decades, in the distribution of biota as a function of depth and position in soils and sediments (see also Section 2.4). Satellite images of other bodies in the solar system also reveal geopatterns that show familiar forms developed in unusual materials and under conditions different from those on Earth.

Most of the natural geopatterns we see are *self-organized*—they emerge spontaneously from local interactions as opposed to being imposed by some outside influence. Patterns that arise from the internal dynamics of landscapes are *autogenic*, a term that encompasses both spatial and temporal variation. Most geopatterns, regardless of scale, are also dynamic—they develop over time and in many cases remain dynamic even when statistically in steady state. Natural geopatterns are also resilient. A kind of "geomorphic natural selection" weeds out unstable patterns and favors those that are resilient in the face of natural forces and fluctuations (see also Section 2.7). Where do these patterns come from? What do they tell us? How can we use them? As old as these questions are, new observational and analytical methods are now available to yield deeper answers to them.

FIGURE 2.5 The cuspate capes of North Carolina, examples of a feature common on sandy coasts, illustrate the emergence of a large-scale structure from the interplay of waves and beach sand. Their characteristic scale is much greater than that of the waves that create them. SOURCE: Jeff Schmaltz, MODIS Rapid Response Team, NASA/GSFC.

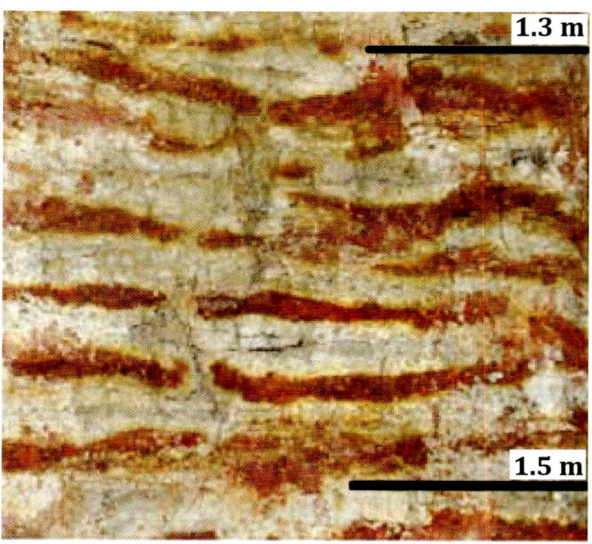

FIGURE 2.6 Biogeochemical reactions can cause pattern formation such as these soil bands. The striping results when biogeochemical processes cause electrons to be transferred to metal oxides containing iron and manganese. The bars represent downward depth in the soil profile. The top bar is at 1.3-meter depth and bottom bar is at 1.5-meter depth; thus the distance between the two bars is 20 centimeters. SOURCE: Fimmen et al. (2007). With kind permission from Springer Science+Business Media.

Grand Challenges in Earth Surface Processes

FIGURE 2.7 The unmistakable human footprint dominates many landscapes. Even someone unfamiliar with central-pivot irrigation systems would have no trouble identifying this pattern as unnatural. Central-pivot irrigation systems tap subsurface groundwater. Each circle can be very large—51 hectares, or 126 acres or more in some places. SOURCE: NASA/GSFC/METI/ERSDAC/JAROS and U.S.-Japan ASTER Science Team.

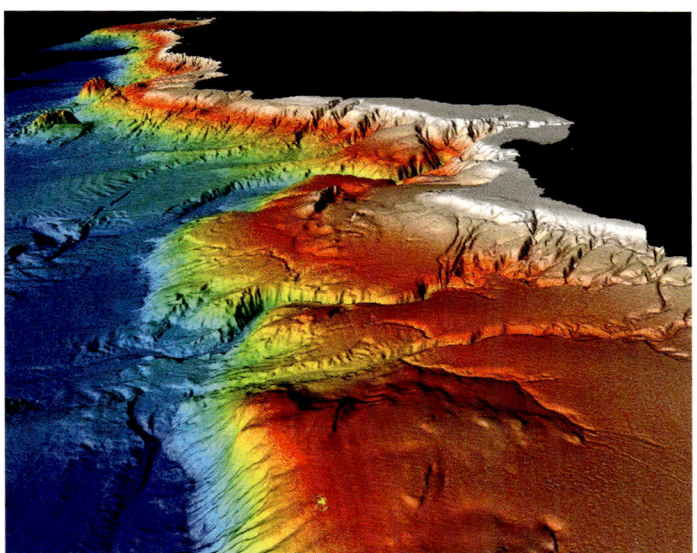

FIGURE 2.8 The submarine landscape off California revealed by high-resolution sonar. SOURCE: Image courtesy of Lincoln Pratson, Duke University.

Pattern formation and self-organization have emerged as interdisciplinary sciences in their own right. Fractal geometry, the study of structures and patterns with non-integer dimensions, arose from an attempt to measure the length of the coastline of Britain. Fractals and their relatives continue to facilitate our understanding of the spatial structure of landscapes. The emergence of complex systems and pattern formation as research areas in physics, mathematics, and other sciences has brought new attention to geopatterns from these communities.

Consider the evenly spaced valleys in locations as disparate as the humid Appalachian Mountains of Pennsylvania, the Coast Ranges of semiarid central California, and the steep flanks of arid Death Valley, southern California (Figure 2.9). How did these valleys form, and

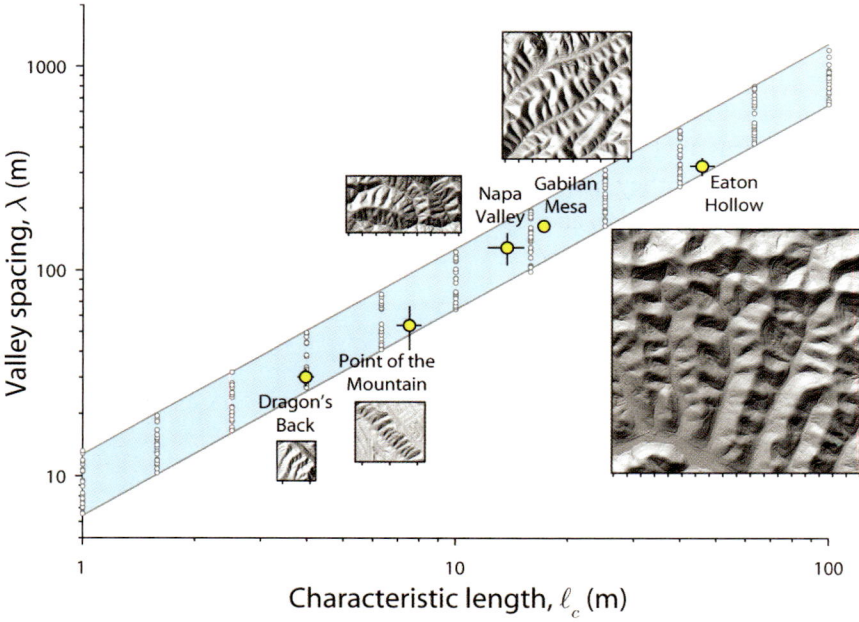

FIGURE 2.9 Comparison of predicted versus observed ridge and valley wavelengths. Valley spacing is plotted against the length at which the time scales for specific types of erosion and stream incision are equal. This length scale is directly proportional to the valley spacing produced in a numerical model of landform evolution (gray points, with linear trend highlighted blue) and to the measured valley spacing at five field sites (yellow points). Insets are shaded relief maps of sections of the sites, with vegetation filtered out of the laser altimetry data to show the underlying topography. Tick spacing is 200 meters. Laser altimetry is from the State of Pennsylvania PAMAP program (Eaton Hollow, Pa.), the State of Utah Automated Geographic Reference Center (Point of the Mountain, Utah), and the National Center for Airborne Laser Mapping (remaining three sites, all in California). SOURCE: Modified from Perron et al. (2009) and used with permission from Macmillan Publishers Ltd, Nature.

what processes control their spacing? Valley spacing appears to be a fundamental, emergent signal that tells us something about the types and rates of erosional and depositional processes that shape the landscape, despite varied climatic settings and geological foundations. A successful theory of landscape evolution must be able to explain such fundamental signals and their variability in space and time. Prior to the availability of light detection and ranging (lidar), however, measuring valley spacing accurately was difficult because of vegetation cover and the relatively low resolution of topographic data (Box 2.1).

Landscape geopatterns provide a template for a broad spectrum of processes on and around the land surface. The subdiscipline of landscape ecology is devoted to the study of spatial patterns in ecosystems, including ties to those in the subsurface (see also Section 2.6). Landslides and river floods, with their economic and human costs, are strongly conditioned by the self-organized geometry of drainage basins and the hydraulic characteristics of near-surface rocks and soils. Intensively patterned soils (vertical profiles and spatial catenas of soil properties), as well, reflect geochemical and physical segregation strongly conditioned by microbial processes (see Section 2.4). Overall, the distribution and intensity of a range of landscape processes, such as erosion and deposition or nutrient uptake and release, are tied to spatial structures in ways that we are only beginning to recognize and understand.

That geopatterns are found throughout the landscape is reason enough to want to know how they arise and what they mean, but there are also pragmatic reasons for studying them. Geopatterns record information about present and past conditions and structure the behavior and response to changes in landscapes and their linked ecosystems. The locations of landslide hazards and river and coastal erosion, for example, are closely tied to spatial patterns in the landscape. New evidence suggests that this is also true of less obvious processes, such as soil biochemistry and nutrient cycling. As we strive to manage natural systems better—to work with nature rather than against it—we ought to understand how natural systems organize themselves and how the resultant patterns may contribute to their resilience (see also Section 2.7).

New observational methods and tools have improved our ability to measure familiar geopatterns, revealed new geopatterns on scales and in places that previously could not be observed (including other planets) and, in some cases, have allowed us to measure rates of pattern evolution. Recent breakthroughs in computational spatial analysis, for instance, using geographic information systems (GIS) to study connectivity, distance-decay functions, shape analysis, edge roughness, and association between patterns show great promise. The need also exists to build stronger linkages between spatial analysis and GIS in the study of Earth surface processes. In addition, advanced tools in modeling and statistical analysis provide powerful new ways to quantify and extract information from the spatial structures of landscapes (see Appendix C). Modeling changes in channel morphology from upstream to downstream can also yield information to understand what controls the rates of change. Below, major opportunities for research in geopatterns are summarized under three broad questions.

How Do Local Interactions Give Rise to Extensive, Organized Landscape Patterns?

The study of pattern formation has already developed into a subdiscipline of its own. Approaches to the origin of landscape patterns span the full range of methods: from classical perturbation and normal-mode analyses to a variety of numerical models including generic evolution equations, coupled systems of partial differential equations, and cellular (for example, coupled map lattice) models. The combination of these new approaches, inexpensive computing, and more rigorous quantitative training of researchers has improved our capabilities for modeling surface geopatterns. Models can now generate plausible representations of most types of landscape patterns, often from relatively simple interactions. However, only in a few cases do we have what might be termed a "standard model"; many of the models proposed so far have enough free parameters to make them difficult to test. Research opportunities include advancing model development, with an eye toward making testable predictions, and field and/or experimental campaigns to discover the critical interactions and to test and constrain models.

What Does Spatial Organization Tell Us About Underlying Processes?

Modeling geopatterns leads naturally to the question of what we can infer from them about underlying processes. For example, we associate tributary (collection) channel patterns with net erosion and distributary (dispersive) patterns with net deposition (Box 2.2). The rapid development of new observational tools such as lidar (Box 2.1) and high-resolution, three-dimensional seismic reflection (Appendix C) provides unusual opportunities to answer new questions. Are aspects of tributary network patterns sensitive to rate and spatial distribution of uplift? Can distributary and shoreline patterns provide quantitative constraints on the relative importance of rivers, waves, and tides in shaping coastlines? What can soil geopatterns tell us about chemical fluxes? Answering these questions would require measurement of local physical, geochemical, and biotic processes and attributes in addition to extant and preserved topography. Thus, relating spatial variation of smaller-scale (hard-to-measure) quantities to larger landscape patterns would be important to reveal underlying explanations. Exploration of these variations using field and laboratory studies is rapidly expanding. An exciting development has been the discovery of process "hot spots"—areas where a high level of activity concentrated in a small location can be identified from relatively simple morphologic measures. Another promising approach is to use topographically based estimates to provide field scientists with a set of reference values for key local variables that serve as a starting template for observation. Both efforts are in their infancy but point to the enormous potential for understanding the Earth surface environment through discovering the quantitative connections between surface patterns and processes.

BOX 2.2
The Natural Mississippi Delta as a Resilient, Self-Organized System

The modern Mississippi Delta illustrates many of the key issues of how humans have affected a natural self-organized system and dramatically reduced its resilience. The motivations for manipulating and stabilizing the Mississippi channel as the coastal economic zone developed are understandable, but their consequences for the delta as a self-maintaining structure have been catastrophic. A delta that has existed since Cretaceous time (between about 65 million and 145 million years ago), has maintained itself through Pleistocene (1.8 million to 10,000 years ago) sea-level changes in excess of 100 meters, and has created most of the modern Louisiana coast over the past 5,000 years has now lost nearly one-third of its total area just since European settlement. Although the land loss is readily quantified in terms of loss of land area over time, it is also interesting to look at in terms of pattern disruption—for example, by comparing the shape and channel network of the current bird's-foot delta (right figure below) with that of a healthy, growing, natural delta created in the Gulf of Mexico from an adjoining distributary but not impacted by human activity (Wax Lake Delta, below left).

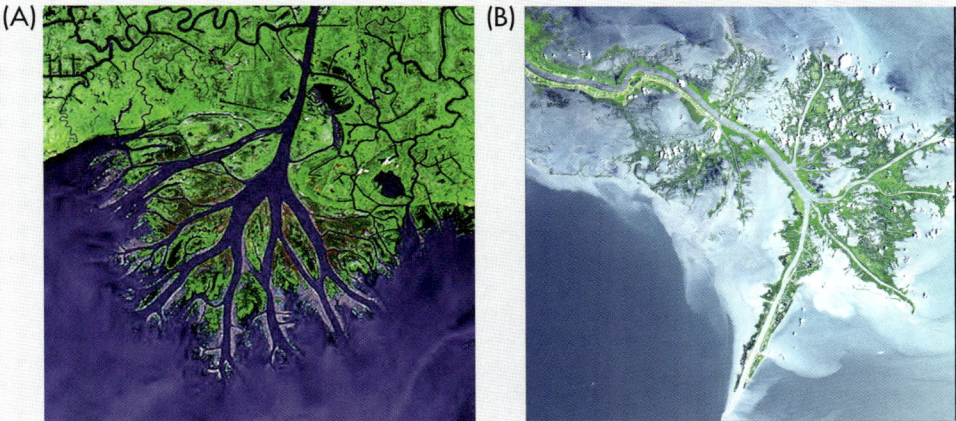

The active, growing Wax Lake Delta (*left*) in contrast to the main bird's-foot Mississippi Delta (*right*), which continues to lose land. SOURCES: (A) LANDSAT 7 WorldWind Geocover 2000. (B) Courtesy of United States Geological Society National Center for EROS and NASA Landsat Project Science Office.

How Can We Use Landscape Patterns to Improve Prediction?

We are currently faced with the problem of forecasting the response of the Earth surface system to changes that include climate, land use, and sea level. The scale and complexity of the system require that we use every means at our disposal to get traction on the problem, realizing that in most cases our predictions, like weather forecasts, will include estimates of uncertainty, reflecting the inherently stochastic nature of the underlying processes. Thus, as we gain a better understanding of the origin and dynamics of patterns on Earth's surface, we must also learn to use them for prediction. For example, a major effort is under way in hydrology to use statistical approaches for the prediction of variables such as streamflow in ungauged basins for which only topography and rainfall data are available. This effort reflects the difficulties in modeling the rainfall-runoff process mechanistically from first principles (see also Section 2.5). Exploiting the pattern of the drainage basin is a key to this effort.

The successful use of patterns for prediction requires that two issues be addressed. First, because any persistent natural pattern must be resilient to some extent in the face of disturbance, the origin and limits of this resilience to change must be addressed (see also Section 2.7). For example, as surface systems are altered by a range of human impacts (Section 2.8), will naturally organized structures be distorted or disrupted and, if so, at what thresholds, and how can these thresholds help us to predict future change? Additionally, prediction of landscape patterns requires expansion into the subsurface (see also Section 2.1). The traditional question of how two-dimensional surface patterns are recorded as three-dimensional subsurface structures to infer three-dimensional rock properties is still important for understanding the flow of subsurface fluids. New opportunities arise from understanding how "fossilized" surface patterns can help us to better understand and predict surface evolution in the present.

2.3 HOW DO LANDSCAPES INFLUENCE AND RECORD CLIMATE AND TECTONICS?

Tectonic plates are hard and dense, and evolve slowly; by comparison, the atmosphere is fluid, thin, and quick to change. One might think that their interaction would be limited to things such as the well-known effects of terrain on local climate. In fact, one of the major advances in the Earth sciences of the last two decades has been to show that the connections between the climate and tectonic systems are far deeper and more subtle than this. The issues include such interdisciplinary concerns as climate variability, the evolution of landscapes and the biological communities (including humans) they host, and the deformation that builds mountains and basins. These diverse processes interact with one another on time scales ranging from millions of years to individual storm and flood events that last hours or weeks, and seismically triggered mass wasting that lasts for seconds but affects the landscape for years to millennia.

Rugged mountain topography reflects the most intense interplay between tectonic uplift and erosional removal of rock material (Figure 2.10). Rates and patterns of deformation in active convergent mountain belts are strongly influenced by the breakdown of rocks and their

Grand Challenges in Earth Surface Processes

FIGURE 2.10 Mountainous topography reflects the competition between tectonic rock uplift and climate-modulated physical and chemical denudation. Climate, topography, denudation, and tectonics are intertwined in a complex manner that operates over a wide range of time and space scales. These dynamic interactions affect civilization on a daily basis. SOURCE: Photo courtesy of Kelin Whipple, Arizona State University.

removal by hillslope collapse, river incision, gullying, and other erosion processes whose actions occur over relatively small scales but cumulatively control the erosion of entire mountains.

The rapid erosion in response to snowmelt and intense rainfall in rugged terrain and the deposition of sediment in neighboring alluvial lowlands in response to flooding are familiar occurrences to those living in mountainous areas. Fault movement, earthquakes, and erosion are also familiar to those living in active tectonic areas. Less well-known is the recent discovery that the long-term rate and pattern of erosion influence the rate and spatial distribution of tectonic motions, both at Earth's surface and in the interior. This recent insight fosters an entirely new set of research questions that can bring together climate scientists, geophysicists, structural geologists, and Earth surface scientists in an effort to understand how processes acting at Earth's surface both influence, and are influenced by, processes within the solid Earth.

The study of Earth's deep interior and much of traditional Earth science cannot be decoupled from study of Earth's surface dynamics. Moreover, the interplay of climate and tectonics produces landforms and deposits that encode the history of climatic and tectonic conditions—a record of essential scientific and societal value. Fundamentally, surface dynamics influences tectonics because topography dictates, in part, the state of stress within the Earth's interior. As a consequence, erosional and depositional modification of topography may produce a direct deformational response. Numerical and physical experiments have shown that a concentration of erosion in one area may trigger the inception of new faults or reactivate old ones. Deposition can also suppress deformation on frontal thrusts and raise base level, thereby reducing erosion rates in the interior of the range. Because rates and patterns of erosion may be climatically controlled, deformation within Earth's lithosphere may also be climatically controlled. Rapid uplift of rocks along New Zealand's Alpine fault, for example, may be ascribed largely to intense precipitation (in excess of 10 meters per year) driven by orographic lifting of moist trade winds impinging on the western flank of the Southern Alps (Figure 2.11; see also Figure 1.2). If precipitation and erosion were distributed more evenly across this mountain range, the rate of vertical motion along the Alpine fault would be a fraction of what we observe today. Biota may also directly or indirectly affect deep-Earth processes and tectonics (Section 2.6) either by altering precipitation patterns (and thereby erosion rates) or by influencing rates of weathering and surface erosion directly.

Some of the most intriguing research questions in the interaction of climate and tectonics center on the relative sensitivity and rates of the numerous feedback mechanisms among climate, topography, ecosystems, physical and chemical denudation, and the deformation of rocks in active, convergent mountain belts. Complex feedback mechanisms involve, for example, orographic enhancement of precipitation, evolution of the temperature field within the deforming interior of the Earth, and the strength and viscosity of rocks. Rapid erosion can significantly increase geothermal gradients, causing elevated temperatures within the crust. These elevated temperatures can weaken rocks, focusing and accelerating deformation, which may in turn increase topographic relief and thus increase chemical and physical erosion rates through a variety of mechanisms, forming a positive feedback loop. Away from active orogenic zones, the lowlands also provide an interesting record of climatic and tectonic events (Box 2.3).

Chemical weathering and biogeochemical cycles influence, and are influenced by, biological processes and physical erosion. Recent findings show that (up to some limit) the faster the physical erosion rates, the faster are the chemical erosion rates. Chemical erosion, in this sense, is loss of mass due to chemical dissolution. This coupling to physical erosion rates establishes an important link to rates of tectonic deformation and

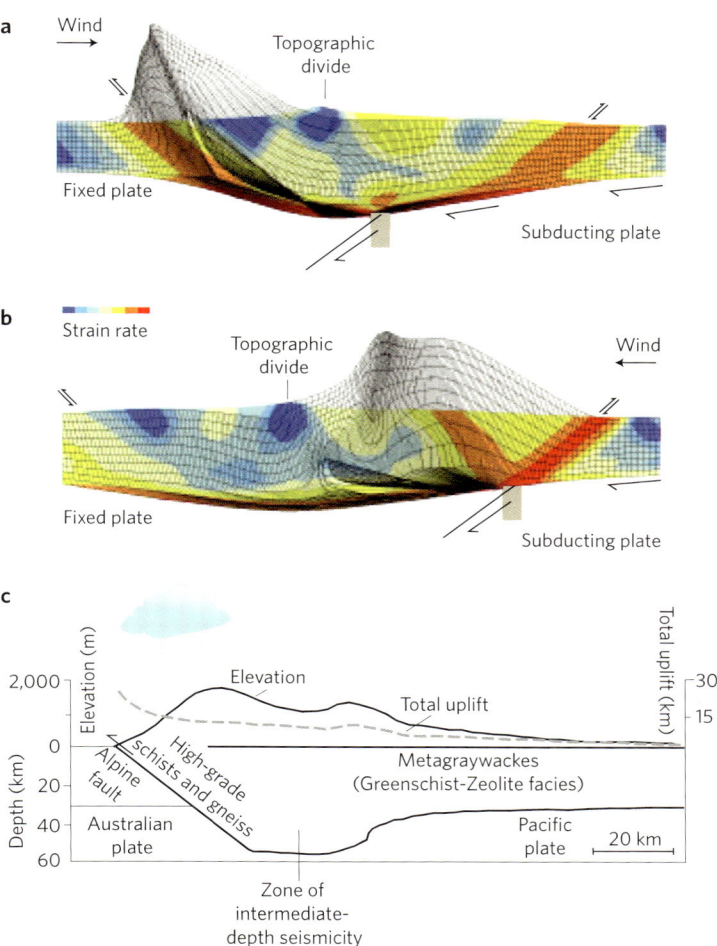

FIGURE 2.11 Unidirectional moisture flux and mountain belt evolution. Top panels (a) and (b) show the results of numerical models aimed at understanding the tectonic response of mountain belts to unidirectional moisture flux. The tectonic convergence velocity and the subduction direction in the models match conditions for the Southern Alps of New Zealand. Red and orange colors represent high strain rates; blue colors indicate low strain rates. In (a), moisture-laden winds arrive from the west (left). The amount of uplift and exhumation (the difference between topography and total uplift) of the mountain belt is indicated by the extension of the gray gridded area (a tracking mesh) above the colored gridded domain (which represents the topographic surface). The uplift and exhumation are focused over an active thrust fault (orange band on the left side of the figure) and west of the topographic divide. In (b), moisture-laden winds arrive from the east (right). Both uplift and exhumation are focused east of the topographic divide. The western thrust fault (same location as in (a)) is nearly inactive. SOURCE: Whipple (2009), modified with permission from Willett (1999); courtesy of Macmillan Publishers Ltd: Nature Geoscience. (c) The observed topography and pattern of total uplift and exhumation in the Southern Alps closely match the numerical experiment shown in (a). SOURCE: Whipple (2009), modified by permission from Koons (1990); courtesy of Macmillan Publishers Ltd: Nature Geoscience.

> **BOX 2.3**
> **Lowland Landscapes and the Record of Climate and Tectonics**
>
> The Earth's lowlands constitute an area that includes most of the modern landscape and the most heavily populated regions of the continents. In these regions, erosion rates have slowed over hundreds of millions years as landscapes have been reduced to generally low elevations with long hillslopes, mantled with relatively deep soils, saprolites, and sedimentary covers. This record of weathering and sedimentation on hillslopes and in lakes and caves across eastern North America, Amazonia, Australia, and Africa is being rapidly unraveled with a combination of geochemical studies; cosmogenic isotopes; detailed geochronology with K-Ar, ^{40}Ar/^{39}Ar, ^{18}O profiles, and paleomagnetism, and modeling of the simultaneous evolution of the chemistry and physics of regolith evolution and transport. The work reveals a rich history of climatic and tectonic events and of recent human occupation. In some parts of these lowland environments, erosion rates by water and wind are high because of the occurrence of sparsely vegetated, erodible sedimentary covers left behind by glaciers and meltwater rivers during glaciation. In others, the low rate of erosion leaves rivers starved of sediment and flowing on bedrock. In regions that have been affected by human colonization and intensive land use during the past several centuries, some of the lowland rivers have been transformed from sediment starved to alluviated (see also Section 2.9). Understanding how these patterns are being created by tectonism, climate change, and sea-level variation over the Late Cenozoic era and the degree to which they have been disrupted by human intervention poses scientific questions that link geological studies of landscape evolution with the need for predictions of the future of landscapes under human and other natural influences (see also Section 2.8).
>
> These continental-scale lowlands also are drained by the world's largest rivers, the vast alluvial lowlands and deltas which host hundreds of millions of people who depend on both the life-sustaining benefits of the floodplain soil and water resources and the environmental hazards that arise from the rivers' hydrologic and geomorphic behavior. Although the generally quiescent tectonic framework of these cratonic regions is the basis for their low topography, low rates of erosion, and thick soil and sedimentary covers, they are nevertheless subject to subtle flexure and faulting that have important consequences for the largest rivers and deltas where small alterations of elevation and gradient affect sediment transport, river courses, and the evolution of deltaic wetlands. Subtle influences of this kind, along with the aforementioned human disruptions pose refractory problems for the management of sustainable valley-floor and coastal regions in most of the major river valleys of the world. Interdisciplinary studies of Earth surface processes provide a geographically and temporally extensive perspective on these problems, which currently are being tackled by local engineering strategies alone. This perspective provides a critical context in which to more fully examine, understand, and address those problems.

establishes a critical negative feedback in the global climate system. Chemical weathering of silicate rocks directly influences global climate by consuming atmospheric CO_2 (see also Section 2.4). Thus, the correlation of chemical weathering with physical erosion establishes one potential mechanism by which tectonics may influence global climate: rapid rock uplift enhances physical erosion, leading to greater chemical weathering and a drawdown of atmospheric CO_2 resulting in a cooler climate. The exact causes and strength of this linkage remain topics of debate. Other linkages between Earth surface processes and global climate that require study involve the sequestration of carbon in the sediments of foreland basins and deltas, the potential for reestablishing carbon storage in soils in the vast cratonic regions that are managed by humans for agriculture and forestry, and the potential for decreases in the carbon reservoir in boreal regions if warming of the soils combined with melt-driven hydrologic and geomorphic disruption causes drainage and oxidation, and methane release.

The linkages among surface processes, tectonics, climate, and landscapes operate at geologic time scales (tens of thousands to millions of years), but have practical implications because they influence the distribution of active faults and their rates of displacement. The record of paleo-earthquakes and deformation patterns preserved in landforms and surficial deposits has been recognized as an essential complement to monitoring active deformation using repeat geodetic surveys, global positioning systems and seismometer networks, and satellite imagery (for example, interferometric synthetic aperture radar [InSAR]) (see Appendix C). These records are particularly useful because historical and instrumental records are too short to fully gauge the seismic hazard potential of earthquakes with great destructive potential and recurrence intervals from 100 to 10,000 years. Research is also being conducted to understand and use the continuous changes in the pattern of strain accommodation along individual faults. Commonly observed differences between slip rates determined over geologic and decadal time scales are driving investigation into the mechanics and history of earthquake clustering, fault growth, and interactions between nearby faults. The only information available for this history is that recorded in landforms and sedimentary deposits (see also Section 2.1).

The availability of new tools and the development of new approaches to the study of this important record of fault activity have opened exciting new opportunities. Airborne and ground-based lidar instruments (Box 2.1; Appendix C) now allow rapid mapping of topography at a spatial resolution and spatial coverage not even imagined a decade ago, greatly facilitating quantitative study of deformed landscapes. Much of this work focuses on the use of slowly eroding landscape features as passive strain markers. High-resolution topographic data are also unlocking a complementary record of tectonic displacements and climatic conditions preserved in the form of actively eroding landscapes (Figure 2.12). New methods of geochronology, including cosmogenic nuclides (Box 1.2), optically stimulated luminescence (OSL), and probabilistic analyses of carbon-14 (^{14}C) data, also afford unprecedented resolution of the age of deposits and landforms, the timing of events, and the rates of deformation and erosion (see also Appendix C).

Several particularly promising lines of research concern the interaction of climate and tectonics. A well-developed understanding exists of the behavior, over geologic time scales, of model systems characterized by simple rules for rock deformation, steady plate

FIGURE 2.12 Dragon's Back Pressure Ridge, Carrizo Plain, Central California. (A) Elevation map derived from digital topography created from airborne laser swath mapping topography (1-meter digital elevation model) (National Center for Airborne Laser Mapping); (B) rock vertical uplift rate along the San Andreas fault. From these data, Hilley and Arrowsmith (2008) demonstrated how landscapes respond to a pulse of uplift. SOURCE: Hilley and Arrowsmith (2008); courtesy of The Geological Society of America.

convergence, subduction of mantle lithosphere, and surface erosion, which is assumed to vary with mean annual rainfall and regional topographic gradient. Although observations are generally consistent with these simple model predictions, we have not yet quantitatively tested the predictions in natural systems—testing these models remains an important research objective and opportunity. In addition, the next generation of models will have to account more completely for the dynamics of the lithosphere and mantle, the complex rheologies of rocks, oblique convergence, the role of sedimentation on the flanks of mountain ranges, and a more quantitative and complete representation of climate relative to its interactions with topography, ecology, and erosional processes and rates.

Four research opportunities emerge as particularly promising toward advancing our understanding of the linkages among climate, surface dynamics, and tectonics: (1) quantification of the role of climate in surface processes; (2) influence of mountain building and surface processes on climate; (3) sedimentation and mountain building (see also Section 2.1); and (4) interactions of surface processes, climate, tectonics, and mantle dynamics.

Quantification of the Role of Climate in Surface Processes

Climate is expected to exert a powerful influence on landforms, weathering, and erosion. However, much remains unknown regarding the causes and effects of these influences, and in some instances, data fail to demonstrate the expected dependences. Thus, accurate, quantitative predictions about the relationships among climate, weathering, topography, erosion, and the rate of sediment delivery to depositional basins are difficult to make. This knowledge gap persists in part because of the complex interplay among the relevant processes and the erosion rate, but it also reflects a dearth of data collected systematically to investigate these relationships. Thus, a high-priority research question remains: How does climate modulate the relationship between topography and erosion rate?

The assumption that erosion rate increases with precipitation (or runoff) is used in most models linking climate, tectonics, and erosion, yet this assumption is not well supported with field data or theory. The relationship between climate and erosion transcends long-term geologic and geodynamic questions. Understanding this relationship is essential to (1) understand the role of climate and climate variability in the form and function of Earth's surface in general; (2) interpret the intensity of tectonic activity and, therefore, seismic hazard; and (3) predict Earth surface response to climate change.

The availability of new tools (for example, OSL, cosmogenic nuclides, isotopic indicators of material sources, flow paths, and residence times; e.g., Box 1.2) creates an opportunity to develop millennial-scale datasets appropriate to quantify the modulation of weathering and erosion by climate. As highlighted in Section 2.1, the sedimentological and isotopic records of paleoclimatic conditions and particulate and chemical fluxes into depositional basins are also valuable information resources to address this research opportunity. Ultimately these datasets can be used to guide and test explicit theories for how climate influences erosion, transport, and deposition processes (see also Section 2.5).

Influence of Mountain Building and Surface Processes on Climate

Surface processes influence local and global climate in numerous but poorly understood ways through their influence on topography, land cover (ecosystems, albedo), soil composition, and soil moisture content. At the largest scale, the rise of mountains alters large-scale atmospheric circulation and its complex interactions with the oceans; mountain belts are both barriers and deflectors and influence the pattern of heat exchange among the Earth, the oceans, and the atmosphere. In addition to affecting circulation patterns and driving orographic precipitation, the rise of major mountain belts can increase climate variability of the atmosphere and oceans over thousands to tens of thousands of kilometers downstream, both through lee-side planetary wave responses (relatively well understood) and through the effects of rugged topography, and complex patterns of albedo and surface energy and water fluxes (less well understood). The influence of land surface-atmosphere interactions (energy and moisture exchanges modulated by topography, biological communities, and soil hydrology) also operates on a more local scale and across all landscapes (not just in the mountains). In addition, oceanic feedback to the atmosphere plays an essential role in climatic response to changes in topography, land cover, and orbital forcing. Ocean feedbacks to changes in large-scale landscapes and climate are also central in determining the impact of sea-level rise on coastal landscapes and understanding geological and paleoclimate records.

Although widely discussed, these feedback mechanisms between landscapes, the ecosystems they host, and climate remain poorly quantified. Indeed, although seen in some models, it has proven difficult to establish from observations whether a significant net feedback of the land surface on climate exists. New modeling tools such as coupled global and regional dynamic vegetation and climate models developed in the past decade, datasets, and statistical analysis methods show promise for resolving this issue. A recent progress report on the U.S. Climate Change Science Program (NRC, 2007b) concludes that inadequate progress has been made on the problem of the potential feedbacks among land-use and land-cover changes, ecosystems, and climate. Most research to date has focused on Earth surface and ecosystem response to climate change. An important opportunity and a challenge lie in the study of the coevolution of climate and Earth surface conditions. Two examples include the carbon budget of permafrost regions and the climate impact of dust aerosols. As areas of permafrost thaw in response to global warming, vast stores of carbon and methane may be released into the atmosphere, establishing a positive feedback (see also Sections 2.4 and 2.7). Dust aerosols have a significant impact on radiative forcing of the climate and on the health of humans and animals. Dust aerosol concentrations depend on soil composition, soil moisture, topography, and land cover in the source region, which in turn depend on current and past physical, chemical, and biological processes over dryland. Dryland ecology is especially vulnerable for projected climatic drying and increasing land use through activities such as grazing. Degradation of dryland ecosystems could further reinforce climatic drying and increase dust aerosols globally.

Sedimentation and Mountain Building

Sedimentation on the flanks of actively deforming mountain ranges significantly influences the rate and pattern of active faulting and erosion rates. Few models, however, have evaluated the role of sedimentation in the evolution of mountain ranges. Insight about this interaction will come in part from analysis of the sedimentary record itself. Sedimentation may generate an extensive record of the history of erosion rates and deformation patterns that can be studied in outcrop and, importantly, imaged seismically in the subsurface (see also Section 2.1). In recent decades, research into the interaction of climate and tectonics has focused almost exclusively on the erosion and exhumation record; new tools have recently

> **BOX 2.4**
> **Thermochronometer Renaissance**
>
> Thermochronometer techniques rely on radioisotopic dating to constrain the thermal histories of rocks and minerals. In the last 5 to 10 years, thermochronometer applications have broadened to include processes at Earth's surface. This breakthrough in Earth surface process-related applications, or the "thermochronometer renaissance," is the result of two developments: (1) new thermochronometer techniques and the refinement of existing techniques such as (U-Th)/He and ^{4}He/^{3}He thermochronometry now enable the reconstruction of near-surface thermal histories of crustal rock materials (temperatures less than ~90° C) in ways that were not previously possible; (2) recent applications of detrital thermochronology from sediments provide new perspectives on spatial and temporal variations in catchment erosion.
>
> Thermochronometer techniques are widely used to quantify landscape denudation rates and magnitudes over time scales of ~10^4 to 10^8 years. The sensitivity of apatite (U-Th)/He data to shallow crustal depths has attracted new interest because subsurface temperatures at, and cooling ages from, these depths are influenced by the shape of the overlying topography. A distribution of cooling ages collected across a mountain range can then be used to infer how the paleotopography and topographic relief appeared at the time of rock sample cooling in that landscape. Recent development of apatite ^{4}He/^{3}He thermochronometry now allows reconstruction of rock cooling histories between ~80° and 20° C. The added information of the cooling history, rather than just the cooling age from the (U-Th)/He technique, provides powerful constraints on the tempo of near-surface denudation. These techniques and extensions of them can, for example, be used to quantify the timing and magnitude of glacial erosion or river incision in a landscape and to measure rates of weathering processes in stable cratonic settings.
>
> Exciting opportunities for applications of thermochronology in the future include (1) addressing catchment erosion processes with detrital thermochronometer samples from river sediments, moraines, hillslopes, and alluvial fans to quantify the original elevation of a sediment source; the distribution of erosion from different geomorphic processes; catchment-wide denudation rates; forest fire frequency and magnitude; or temporal variations in denudation; (2) application of (U-Th)/He and ^{4}He/^{3}He thermochronology to mineral phases sensitive to very low, near-surface temperature histories to date weathering processes and/or reconstruct climate change; and (3) high-density sampling of bedrock across mountain ranges to reconstruct paleo-topography over million-year time scales.

become available that offer an opportunity for new studies of the sedimentary record of mountain building. These tools include cosmogenic nuclides (Box 1.2), low-temperature thermochronology and detrital thermochronology (Box 2.4), and three-dimensional imaging of subsurface sedimentary structures (Appendix C). Three-dimensional imaging of subsurface horizons allows extraction of a time sequence of high-resolution paleo-land surfaces—a direct record of landscape evolution and a resource that has hardly been tapped to date (see also Section 2.1).

Interactions Between Surface Processes, Climate, Tectonics, and Mantle Dynamics

Study of the interactions between climate, topography, erosion, and mantle dynamics is an exciting, novel area of interdisciplinary research involving atmospheric scientists, geodynamicists, geophysicists, and Earth surface scientists. This research is at the frontier of problems in mantle tomography, modeling of mantle convection, lithospheric geodynamics, orographic precipitation, and tectonic, topographic, and climatic controls on erosion and deposition. Geophysicists can now image the current buoyancy structure of the mantle at moderate resolution. This information can be used to seed models of mantle convection that can then be run either forward or backward in time to predict, or retrodict, motion in the mantle over space and time. These models can also be used to generate time series of expected patterns of rock uplift and subsidence associated with mantle convection. An exciting frontier of research lies in the coupling of surface process models to mantle convection models to explore potential feedback mechanisms: Can climate, through the agencies of physical and chemical denudation, sediment transport, and deposition influence convection in the mantle? Equally important will be observational studies to test model predictions and refine models.

Opportunities for Interdisciplinary Collaboration

New opportunities for advancing the theory of weathering and erosion are arising through nascent interactions among atmospheric scientists and Earth surface scientists who traditionally have worked independently. One interdisciplinary research need in the area of interacting landscapes, climate, and tectonics is for a more sophisticated understanding of the controls on the frequency, intensity, and duration of rainfall at the regional scale as a function of local topography, land cover, and global climate. Meeting this need is a major challenge for atmospheric science and is vital to quantitative predictions of the influence of climate on landscapes, ecosystem functioning, erosion, sediment production, transport, and deposition.

At a larger scale, what might be the connections between landscapes and climate due to the topographic and land-cover influences on atmospheric conditions and circulation patterns? How do we build coevolving topography, landscape, and tectonics models? These and other questions will become accessible through direct collaborations among climate scientists and Earth surface process scientists. Insights and tools from several other grand challenges (for example, see Sections 2.1, 2.4, 2.5, and 2.6) will also be required to do so.

Recent research progress sets the stage for a transformation in scientific understanding of the impact of climate and climate change on the Earth's surface. Hydrologists are making significant progress on predicting soil moisture patterns and runoff in response to known rainfall inputs. Collaboration between hydrologists and other Earth surface scientists is beginning to shed light on the coevolution of topography, hydrology, and biological communities (see Section 2.6). Independently, atmospheric scientists and ecologists are making significant progress on quantifying the influence of land surface-atmosphere interactions (energy and moisture exchanges modulated by topography and biological communities) on regional climate (NRC, 2007b). However, further progress is required on studies of the feedback mechanisms between, and the coevolution of, the Earth's surface, its ecosystems, and climate. In addition, developing a theoretical understanding of spatial and temporal patterns of rainfall and how these patterns are controlled by regional climate, topography, and land cover remains a formidable challenge. For example, although climate models can predict differences in mean annual precipitation between the Cascades, the Rocky Mountains, and the Himalayas, neither sufficient data on the differences in storm rainfall statistics between these sites nor theories or models capable of explaining them exist. Computational initiatives in regional climate modeling, wind and wave energy fields, orographic precipitation, and landscape evolution (for example, the Community Surface Dynamics Modeling System [CSDMS][3]) have begun to address some of these issues. Orographic and land surface effects on rainfall are particularly challenging problems, as is predicting seasonal dynamics of rainfall in mountain regions; these effects are important but traditionally have received little attention in climate science.

2.4 HOW DOES THE BIOGEOCHEMICAL REACTOR OF THE EARTH'S SURFACE RESPOND TO AND SHAPE LANDSCAPES FROM LOCAL TO GLOBAL SCALES?

If Earth's surface consisted solely of outcropping bedrock, the surface area available to anchor and nourish life would be a small fraction of what we see today. Instead, in the near-surface environment, rocks disaggregate into particles and react to form new minerals during the formation of soils, which retain some of the solutes for time scales that support plant growth and other biogeochemical processes. The weathering and eroding landscape varies both chemically and physically over space with strong patterns that reflect topography, lithology, biota, and climate; these changes occur over time in ways that we cannot yet predict quantitatively. Importantly, such bedrock weathering processes contribute to landscape evolution, influence biogeochemical fluxes, and impact regional climate. As landscapes evolve, biota play an active role in retaining some of the soluble elements, serving to anchor existing soil on hillsides and to accelerate soil formation. *The breakdown of bedrock—a major factor in Earth surface processes—is among the least understood of the important geological processes.*

The transformation of rock into small and large particles is driven by stresses associated with tectonics, chemical reactions, topography, salt and ice growth, mineral swelling, biotic activity, and thermal fluctuations. This transformation process can occur in an instant,

[3] http://csdms.colorado.edu/wiki/index.php/Main_Page.

such as when a tree, rooted into underlying rock, suddenly falls over, pulling up roots and underlying rock. The destruction can also be slow as waters percolate gradually through the bedrock and progressively leach mobile elements to create openings. A comprehensive theory of how these processes interact and how the rates of chemical weathering and soil formation are related to environmental conditions is lacking. The near-surface fracturing of rock and the transformation of rock into particles are key steps in converting mechanically strong, low-permeability bedrock into an erodible, hydrologically active material. Particles derived from rock become the soil covers of hillslopes and the sediments of rivers, floodplains, deltas, and beaches. Thus, the process involved in transforming rock into soil influences all aspects of Earth's surface dynamics.

Soil formation is not, however, only of academic interest. Our food comes from plants grown in soil. The rapid rate of soil erosion due to land use relative to the slow rate of transformation of rock into soil endangers soil resources worldwide (Box 1.1). The fate of soils, the base of agriculture, is of great concern.

The chemical weathering of rocks and soils affects climate, the chemistry of groundwater and rivers, the strength of rocks and erodibility of landscapes, the availability of nutrients in soils, the fate of anthropogenic contaminants, and the properties of the ecosystems that cover Earth's surface. Even the long-term evolution of the atmosphere is driven by these biogeochemical processes, coupling the fluid envelopes to the solid Earth in feedback loops that, in turn, govern climate. Soil, Earth, and life scientists of all backgrounds are crossing disciplinary boundaries to develop a more integrated picture of Earth's surface down to and including altered bedrock. This new aspect of research on Earth surface processes emphasizes the biogeochemical cycles of all the elements, at all depths, at human-impacted and pristine localities, over all relevant time scales.

Perhaps the most significant of these biogeochemical reactions involves carbon. Throughout the history of Earth, CO_2 emitted to the atmosphere has combined with water to return to Earth as carbonic acid (H_2CO_3) in precipitation. This acid dissolves minerals and then travels in rivers to the sea as bicarbonate, where it precipitates with cations as sediment. Hence, this chemical reaction removes CO_2 from the atmosphere and, on geologic time scales, plays a major role in mediating atmospheric levels of CO_2 and consequently climate. This series of chemical processes links volcanic activity (releasing CO_2 to the atmosphere), mountain building and chemical weathering (more fresh rock and faster CO_2 removal from the atmosphere), and climate (temperature linked to concentration of CO_2).

These processes can proceed in the absence of life, but on Earth, biotic processes strongly interact with the carbon cycle. The litter and decomposed organic matter of soils contains about four times more carbon than in the entire terrestrial biosphere (Sposito, 2008). Litter and soil carbon can be sequestered through accumulation or burial for thousands of years only to be rapidly released back to the atmosphere through disturbances such as agriculture, fire, or thawing of permafrost. Hence, the soil reservoir and vegetation influence short-term carbon dynamics. Considerable research is under way on the dynamics of this soil reservoir of carbon, particularly in the light of how it will change with changing climate and land use. It is a challenging problem that requires understanding the interactions of such factors as vegetation, insect infestation, hydrology, fire, deforestation, agricultural practices, and feedbacks with global warming. Current models disagree on whether soils will be a source or a sink of carbon over the next 100 years, and such soil carbon behavior will be affected drastically by land management practices. The inevitable release of methane associated with the thawing of frozen ground in northern latitudes is also of special concern (see Section 2.7).

Just as we cannot accurately model the carbon cycle, we similarly do not have the tools to project the fluxes of the major nutrients nitrogen and phosphorus as a function of climate, land use, lithology, or other variables. Little research has focused on the effects of different crops on the nitrogen-to-phosphorus ratios in runoff. The implications of widespread increases in nitrogen fluxes in precipitation to landscapes far removed from the source of the nitrogen are also unknown. Furthermore, biogeochemical cycling within the Earth's surface is strongly affected by rapid invasions of species as well as slower ecosystem responses to changes in climate or land use. How do ecosystem changes, often driven by climate change, interact with nutrient availability?

Of further interest is the fact that more than 30 bioessential elements are also extracted from solid Earth materials by biota, often biogeochemically coupled with the cycling of carbon, nitrogen, and phosphorus. For example, molybdenum acts as the most important catalytic center in fixing gaseous nitrogen into the nitrogen species used by biota. Other micronutrients play important roles in the health of ecosystems and in human beings in general. It has been suggested that an increase in the availability of micronutrients might improve human health globally.

Understanding the biogeochemical cycling of elements needed by biota in the present and the past will allow better understanding of how much human intervention through land-cover change will affect the natural balance of biogeochemical cycles (see also Sections 2.6 and 2.8). New technological advances that allow measurement of the isotopic signature of many of these biogeochemical cycles are revolutionizing our understanding of how metal nutrients are cycled by biota. Changes in metal content of soils and sediments over geologic time are documented in the isotopic signatures of elements. Studies of the mobility of minor and trace elements in the modern and ancient Critical Zone may illuminate responses to environmental and climate change.

The prediction of biogeochemical evolution of water from its arrival as precipitation striking the canopy of vegetation through its reaction with soil and rock to its discharge via rivers to the sea is one of the great challenges in Earth surface processes (see also Chapter 1).

> **BOX 2.5**
> **Critical Zone Observatories and Critical Zone Exploration Network**
>
> The Critical Zone Observatories (CZOs) are natural laboratories built around watersheds or groups of watersheds that are investigated by interdisciplinary teams of National Science Foundation (NSF)-funded ecologists, geochemists, geologists, geomorphologists, hydrologists, and soil scientists using field, laboratory, and modeling approaches. Based on the initiatives put forward in a document "Frontiers in Exploration of the Critical Zone" (Brantley et al., 2006) as well as discussions with the Consortium of Universities for Advancement of Hydrologic Science, Inc. (CUAHSI),[a] the first three CZOs were funded from the Division of Earth Sciences (EAR) within the NSF Geosciences Directorate. These three CZOs received five-year grants beginning in 2007 to investigate chemical, physical, and biological processes within the natural laboratories. The funding is being used to emplace instrumentation, collect data, and use models to answer forefront questions at the crossroads of hydrology, geomorphology, soil science, geophysics, ecology, and geochemistry. Implicit to the goals of the CZO program is recognition of the contribution that Earth scientists make toward understanding the Critical Zone.
>
> The three CZOs represent sites that vary in some attributes; for example, two sites lie on granitic material (one in the Sierra foothills of California and one in Colorado) and one on shale (in Pennsylvania). Furthermore, each of the three CZOs is designed around a different set of questions: the Southern Sierra CZO is studying how a change from snow- to rain-dominated climate is impacting hydrology and biogeochemistry in the CZO; the Boulder Creek CZO is studying the influence of differing erosion histories on Critical Zone architecture and function; the Susquehanna-Shale Hills CZO is measuring water, energy, and solute budgets from the water table through the atmospheric boundary layer with particular focus on rates of regolith formation. Three new CZOs have been funded in 2009.
>
> Although each CZO is different, integrative modeling efforts are going on at both the intra-site and the cross-site levels. For example, at the Sierra CZO, models are addressing how the snow-rain transition affects the Critical Zone across an elevation gradient. Similarly, researchers at all three CZOs are using cosmogenic nuclides and models to quantify the rates and mechanisms of the bedrock-to-regolith transformation. As part of this latter initiative, data from the CZOs are being compared to data from seed sites within the Critical Zone Exploration Network (CZEN) as well as to the six satellite sites associated with the Pennsylvania CZO to investigate how variations in temperature and precipitation affect regolith formation

Aspects of this challenge have been addressed—though as yet incompletely. The Critical Zone Observatories (CZOs) will add considerably to this knowledge (Box 2.5).

Water chemistry can be monitored intensively, but the chemical and material properties of the soil and bedrock that influence this chemistry remain technically challenging to document, especially at larger spatial scales. Models for evolution of soil chemistry have been developed but are limited by our lack of knowledge with respect to (1) the kinetics of chemical reactions in nature; (2) fundamental thermodynamic equilibrium constants;

on shale and granite. Projects that span all three CZOs also use fluorescence spectroscopy to characterize dissolved organic matter in surface waters and new instrumentation to quantify water fluxes through efforts facilitated by CUAHSI.

To coordinate the CZOs, the group holds an annual all-CZO meeting. In addition, occasional workshops are held. For example, a small workshop promoted by the Boulder CZO led to the ongoing effort to characterize dissolved organic matter in the CZOs with monthly samples. In addition, a National Steering Committee has been constituted with members from hydrology, geomorphology, and geochemistry drawn from academia and government agencies. The CZOs encourage outside collaborators to use the facilities with funding available in existing NSF programs. Such collaborations are initiated by contacting the principal investigator of each CZO to use instrumentation and data. To foster collaborations, each CZO seeks to establish infrastructure, data sharing, and models that attract researchers in areas of interest. For example, the rich sets of hydrological, geochemical, geomorphological, and ecological data that are collected at each CZO generally do not yet include many of the new isotopic tracers although use of these methods is anticipated in future research at the observatories. Furthermore, the rapid growth in molecular biological and hydrological sensor techniques also represents an opportunity for geobiologists and hydrologists to build on the CZO efforts. CZO investigators are developing mechanisms to share data for the sites online. In addition, Critical Zone scientists in Europe, Australia, and China are developing collaborative projects with CZO scientists as observatories are created abroad.

The CZOs were established separate from the Long Term Ecological Research (LTER) Network and the emerging National Ecological Observatory Network (NEON) in recognition that questions in Earth science related to the Critical Zone are often best addressed at sites chosen to study specific Earth surface processes. LTER sites provide detailed geochemical and biological observations, but often provide less focused geochemical, hydrological, and geophysical data compared to the CZOs (see also Chapter 3). Other proposed networks are under discussion through efforts of the initiative known as the WATERS Network (WATer and Environmental Research Systems Network), currently pursued by the science community and the Engineering; Geosciences; and the Social, Behavioral, and Economic Sciences Directorates in partnership at NSF.

---

[a] http://www.cuahsi.org/hos.html.

---

(3) the influence of biota; (4) the coupling between chemical and physical processes; and (5) the reactivities of materials ranging from fractured bedrock to nanoparticles.

Numerical models have also been developed that couple hydrology to simplified chemical reactions. However, such models have not yielded mechanistic (quantitative and process-based) predictions of first-order observations such as the discharge-dependent concentration of specific elements in river runoff—an achievable goal. The problem also remains of predicting the pathways and residence times of water in the subsurface en route to the channel and the chemical evolution of stream water as it interacts with its bed, floodplain, and biota. Figure 2.13 summarizes the various scales of these processes. Water in a stream (2.13A) integrates waters that have traveled through soil and bedrock (2.13B) dissolving and exchanging elements with the evolving soil and weathered bedrock. Within the soil and rock, mineralogical changes document the effects of chemical reactions at the scale of hand samples, mineral grains (2.13C and D), and the mineral interface (2.13E).

These chemical and physical processes produce a weathered residue that is typically strongly structured vertically across the landscape. Soils of varying thickness and properties typically overlie weathered bedrock. Soils are distinguished by physical and chemical differentiation into distinct layers or horizons—all soil classification systems are based on such horizons. In environments of little physical erosion (typically flat surfaces), soils progressively develop and the changes can be used to assign relative ages of landscapes. On sloping landscapes where soils are eroded and transported, soil properties may become essentially time independent. Greater understanding of soil-forming processes is needed to predict the vertical structure of soil and the time evolution of its properties. Such understanding may even prove useful for planetary studies. For example, some researchers now believe that Mars passed through an early, relatively wet phase of clay formation followed by drying and salt accumulation. Future missions to Mars will examine soil profiles for clues about Mars climate history and the prospect of past life.

The development of cosmogenic nuclide methods (Box 1.2) for calculating lowering rates (denudation) of landscapes has led to new observations relating chemical and physical erosion to climate and tectonics. Early results have been surprising and have profound implications. In many cases, researchers have detected little of the expected climatic (temperature and precipitation) influence on chemical weathering rates. Instead several studies have found that rates of chemical erosion (weathering) vary directly with the rate of physical erosion: the greater the physical erosion rate, the greater is the chemical erosion rate. This finding is also detected in data from river monitoring of sediment and dissolved load. One inference is that more rapid physical erosion removes weathered detritus, exposing fresher surfaces for faster chemical reactions, and that this influence is stronger than that of temperature and water flux. Since physical erosion rates are observed to increase with uplift rate, these findings have important implications for modeling the coevolution of tectonics, topography, and climate (see also Section 2.3). These intriguing surveys are now generating investigations to produce mechanistic models.

The transformation of rock to erodible debris and the shedding of mass through dissolution by subsurface flow influence many processes. Advances in monitoring technologies, geochemical analyses and models, shallow subsurface geophysics, geobiology, nanogeoscience, and rock mechanics will help move the field of Earth surface processes from correlation to explanation and from mapping to prediction.

Grand Challenges in Earth Surface Processes

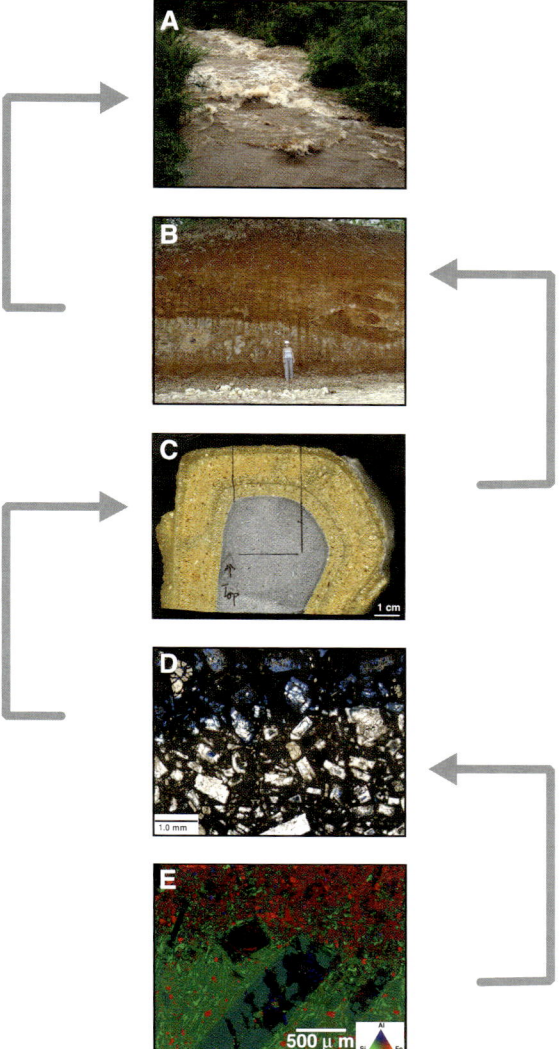

FIGURE 2.13 From the watershed to mineral surface, the biogeochemical evolution of landscapes is driven by processes occurring at many scales. Fluxes of solutes and sediments in flood-stage rivers after Hurricane Dean on Basse Terre Island, Guadeloupe (A), are controlled by soil profiles developed on volcanic debris throughout the watershed (B). Weathering reactions at the clast scale, as shown here to be varying from rock (gray) to soil material (tan) (C), control the chemistry of pore fluids. Dissolution occurs at the rock-rind interface (D), where white, gray, and blue regions are crystallites, glass grains, and porosity, respectively. Finally, the chemistry and topography of individual grain surfaces document the effects of dissolution at the micron scale (E). Photographs A through D derive from the Critical Zone Exploration Network site on Guadeloupe. Photograph E shows basalt weathering in Costa Rica. SOURCES: (A) Courtesy of Jerome Gaillardet, Institut de Physique du Globe de Paris, France. (B) through (E) Courtesy of Peter Sak, Dickinson College © 2009.

2.5 WHAT ARE THE TRANSPORT LAWS THAT GOVERN THE EVOLUTION OF THE EARTH'S SURFACE?

Do the shapes and organized patterns of landscapes reveal the processes that formed them (Figure 2.14)? Are particular landscapes more likely to change rapidly as the Earth warms and as human activity expands across Earth's surface? Complementing the research challenges posed in other sections of this chapter for these kinds of fundamental questions (see also, for example, Sections 2.1 and 2.8) is the need for a mechanistic understanding of processes that link climate, hydrology, geology, biota, land use, topography, and erosion rates. To tackle this challenge we need to discover, quantify, test, and apply laws that define the rates of processes shaping the Earth's surface.

All fields of study search for rules or laws that relate cause and effect, action and reaction, or process and form. In the sciences we typically seek mathematical expressions that can be tested and applied widely. In the physical sciences these commonly are the constitutive relationships that form the foundation of the prediction of deformation, flow, and transport of various materials. Some disciplines have made great strides in this endeavor and can focus, therefore, on the richness of behavior that these laws predict. Such predictions can be as simple as the oscillation of a pendulum or can require linking many laws to explain a complex interacting system, as in global climate modeling or ocean circulation simulations.

A simplified set of mathematical expressions for mass conservation applicable to landscape evolution modeling spotlights the critical need for laws in Earth surface processes. Landscape evolution involves the coupled evolution of two layers (Figure 2.15): a bedrock layer with thickness b, which is the vertical height of the bedrock surface above some datum (usually sea level), overlain by a layer of transportable debris (soil and/or sediment) with vertical thickness h. The elevation of the land surface above the datum is $z = b + h$. The coupled mass conservation equations are

$$\frac{\partial z}{\partial t} = \frac{\partial b}{\partial t} + \frac{\partial h}{\partial t}$$

$$\frac{\partial b}{\partial t} = U - E$$

$$\frac{\partial h}{\partial t} = \frac{\rho_r}{\rho_s} E - \nabla \cdot \mathbf{q_s}$$

The evolution of the land surface depends on the thickness change of the bedrock with time (t) above the datum and the thickness change of the soil-sediment mantle. Bedrock thickness *(b)* is a balance between U, the rate of bedrock uplift (or, if negative, subsidence),

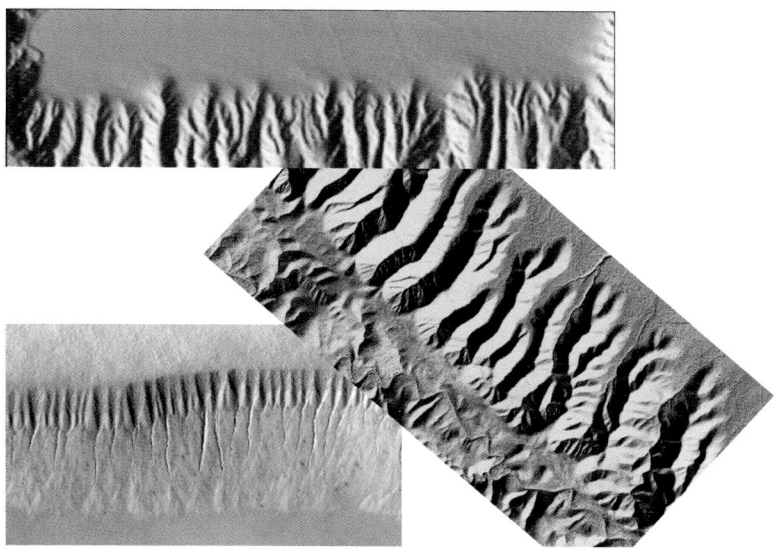

FIGURE 2.14 Three landscapes showing ridge and valley topography. The marine image (*top*; continental slope off Virginia) is derived from echo-sounder data of water depths of 200 to 1,000 meters. The San Andreas fault image (*middle*) is derived from airborne laser swath mapping from http://www.opentopography.org, and the Mars image (*bottom*; NASA/JPL/MSSS) is a photograph from the Mars Observer Camera. SOURCE: (*top*) Reprinted from Mitchell and Huthnance (2007) with permission from Elsevier; (*middle*) courtesy of Ramon Arrowsmith (Arizona State University) through http://www.opentopography.org; and (*bottom*) courtesy of NASA/JPL/MSSS.

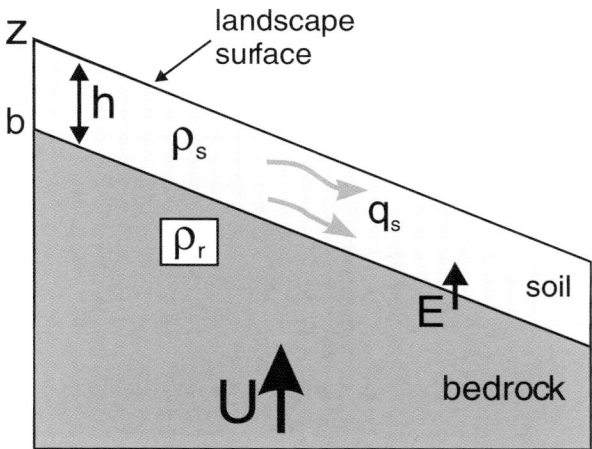

FIGURE 2.15 Profile through a hillslope, showing the uplift of bedrock above some datum (U), its conversion to soil (E), the thickness of soil, h, and the transport of it (q_s) all influencing the land surface height above the datum, z. SOURCE: Modified from Roering (2008); courtesy of The Geological Society of America.

and E, the rate of conversion of bedrock to transportable sediment. The transportable debris (soil or sediment) thickness *(h)* is a balance between the conversion rate of bedrock to transportable debris and the spatial gradient of the volume flux of transportable debris, $\mathbf{q}_s$ (a vector quantity). The parameters ρ_r and ρ_s are the bulk densities of rock and soil or sediment (Figure 2.15). For simplicity, we neglect chemical processes and assume that conversion of bedrock to sediment is strictly mechanical.

This equation is equally applicable to the practical problem of predicting the production and routing of sediment through watersheds (although the uplift term is usually ignored). In this compact form, the influences of climate, geology, land use, and biota are embedded in the E (bedrock erosion) and $\mathbf{q}_s$ (sediment transport) terms.

To solve these equations, we need to know or be able to predict or retrodict the rate of uplift or subsidence of the landscape and to have mathematical expressions for bedrock erosion (or conversion of bedrock to soil) and sediment transport and deposition (Figure 2.16; Box 1.1). Comparison with any real landscape also requires that the first mass conservation equation be integrated over some time period; therefore, the documentation of landscape evolution and the study of landscape history are essential for validating attempts to explain landscape evolution quantitatively. Here, the focus is on the erosion and transport processes.

Developing Mechanistic, Field-Tested Mathematical Expressions

At present we lack mechanistic, field-tested mathematical expressions for most of the processes that shape Earth's surface (Figure 2.16). This gap represents a great opportunity and an enormous need for future research. In the Earth sciences, this research challenge is similar to the need to develop friction laws for faults or temperature- and pressure-dependent viscosity models for mantle convection and glacier flow. The large knowledge gap in Earth surface processes is particularly challenging because it requires dealing with complex materials (for example, soil, organic matter, and bedrock) that deform, flow, and change properties over a wide range of scales of time (seconds to tens of thousands of years) and space (microns to kilometers) as various forces and transport processes (for example, water, wind, ice, and biota) act upon them (see also Sections 2.3 and 2.4). Some models from data collected at field sites are beginning to show the implications of various transport laws on the form of the land surface, taking into account features such as bioturbation and soil thickness variations (Figure 2.17).

The mass conservation equation has no explicit time scale, so the erosion and sediment transport laws could be written to apply either to instantaneous movement or for longer time periods during which short-term variations—such as individual storms, dissolution, or particle motions—are averaged. The latter are particularly appropriate when the landforms of interest, such as drainage basins and mountain ranges, evolve over tens of thousands

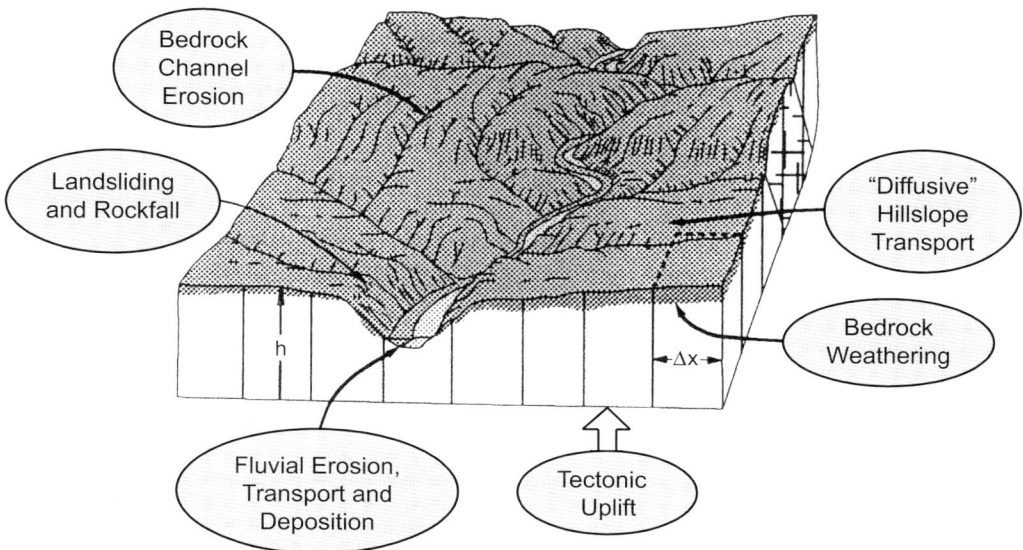

FIGURE 2.16 An example of a set of processes influencing landscape evolution, each requiring distinct transport laws, based on observations. The uplift of the bedrock is balanced by various erosion processes. Deep-seated landslides may develop (for example, along bedding planes). Burrowing and tree throw rip up the bedrock, producing soil and causing it to move downslope, where it accumulates over time. Concentration of subsurface water runoff elevates pore pressures, leading to destabilization of the debris and landsliding. The landslide is mobilized as a debris flow that sweeps away stored sediment and scours the bedrock, cutting the valley. Where the debris flow reaches the mainstem it deposits, and river flows pick up the sediment and remove it, causing wear of the underlying bedrock and cutting the river valley. Hence, at a minimum, to describe quantitatively the landscape evolution pictured here would require the transport or erosion laws for (1) deep-seated landsliding, (2) soil production, (3) soil transport, (4) soil landsliding, (5) debris flow runout and bedrock incision, (6) river sediment transport, and (7) river incision into bedrock. Inclusion of the effects of weathering and chemical erosion would require other transport laws. SOURCE: Tucker and Slingerland (1994); reproduced with permission of the American Geophysical Union.

to millions of years. Transport laws applicable at these longer time scales, or *geomorphic transport laws*, should have parameters that are physically meaningful and can be evaluated with field measurements. Ideally, such parameters would allow us to conduct scaled experiments in the laboratory or to apply transport laws in new settings, such as Mars or Titan.

Significant progress has been made recently in developing and applying geomorphic transport laws for soil transport and river incision into bedrock. These laws are used to explore many important issues, including the relationships among tectonics, climate, and

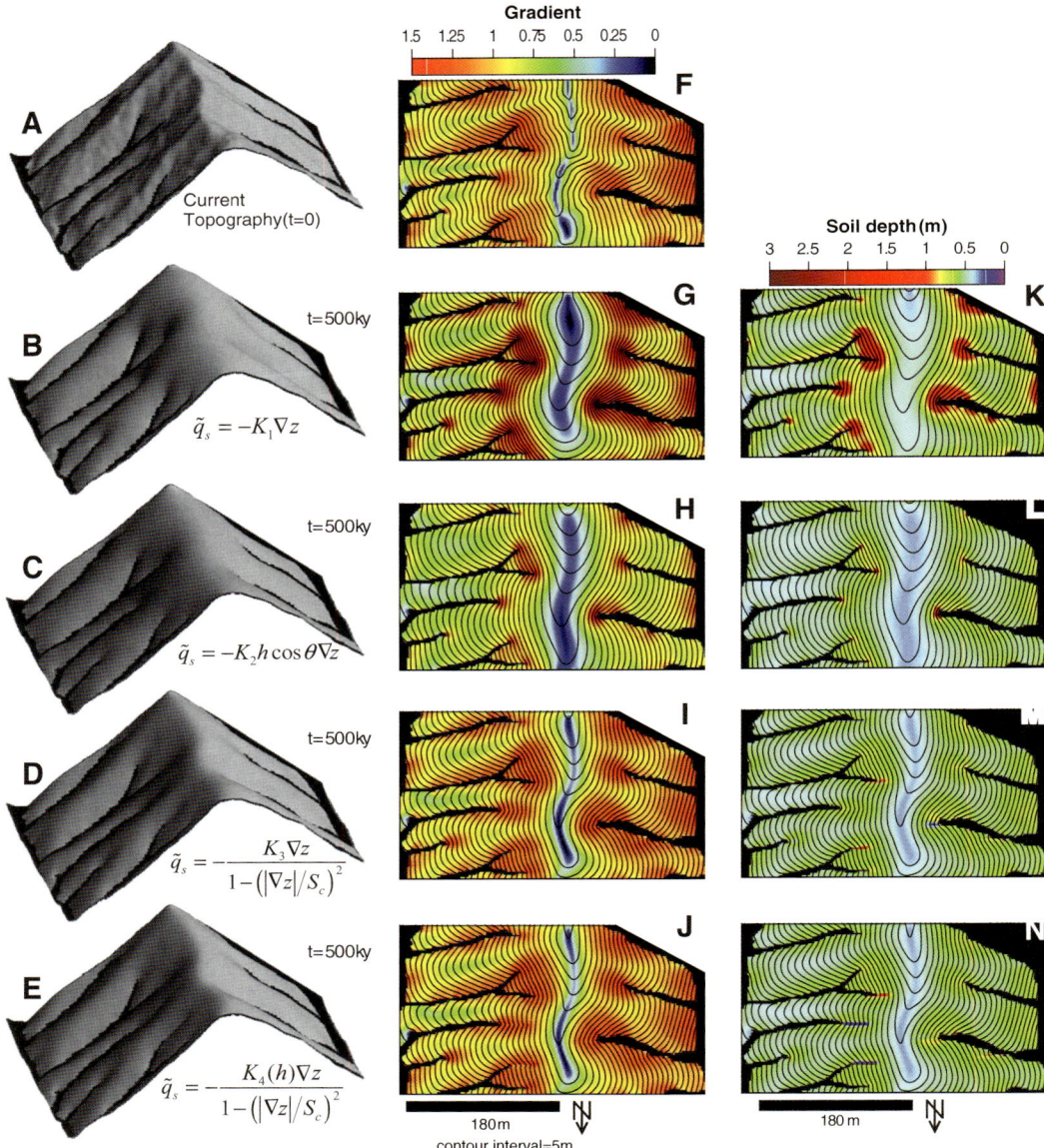

FIGURE 2.17 Comparison of simulation surfaces with current topography. (A)-(E) show perspective-view, shaded relief images of current and modeled topography. Modeled surfaces reflect 500,000 years of evolution via a set of calibrated parameters. (F)-(J) show spatial variation of hillslope gradient for current and modeled surfaces. The current surface is pockmarked due to bioturbation and data errors, whereas the modeled surfaces are uniformly smooth because of the continuum assumption used. (K)-(N) show spatial variation of simulated soil depth for the four transport models. Each model predicts thin soils near the ridge top and thicker soils along side slopes. SOURCE: Roering (2008); courtesy of The Geological Society of America.

topography; the response time of landscapes to changes in uplift; and the processes that control the heights and gradients of hillslopes. These transport laws now are being revised and expanded to include the effects of climate, grain size, and sediment supply, and discoveries made with these laws are being applied in hydrology, geotechnical engineering, and geophysics.

In contrast to these examples of progress, we still lack geomorphic transport laws for such fundamental processes as landsliding, overland flow erosion, glacial erosion, chemical erosion, wind erosion, and transport and deposition of flocculated mud. No laws exist to explain or predict the erosion of bedrock-dominated landscapes or what controls the size of sediment shed from hillslopes into rivers. The breakdown of bedrock into erodible debris, the first step in hillslope erosion, is also poorly understood. Recent studies suggest that as landscapes erode, they create strongly curved surfaces that generate sufficient force to crack bedrock, making it more likely to be eroded (Figure 2.18). Hence, bedrock strength decay, topography, and erosion might coevolve.

A particular mathematical difficulty arises when models are developed to incorporate physical, chemical, and biological processes because the thermodynamic and kinetic equations describing chemical and biological reactions often are highly nonlinear and operate over vastly different time scales. Furthermore, the rates of geochemical and biochemical reactions appear to differ between laboratory and field settings, and we therefore lack confidence in extrapolating geochemical kinetics laws from one system to another. As a consequence, only a few models incorporate sediment transport, geochemical, and biochemical transformations (see also Section 2.4 for discussion of the role of chemical weathering and connections to biogeochemical cycles).

Although it represents a difficult problem, developing transport laws for these processes is now possible. New dating methods enable us to determine rates of processes and their spatial variation across landscapes. Reactive transport codes prove useful in extracting the laboratory-field discrepancy for extrapolating geochemical kinetics. High-resolution topographic surveys made possible from airborne and ground-based lidar surveys can resolve landforms over large areas at sufficiently fine scale to link process and form mechanistically. NSF has also supported NCALM to provide research-grade topographic data, to advance the technology, and to provide education and training for students (see also Box 2.1). Innovative field instrumentation enables us to monitor processes directly and motivate and test conceptual models (Figure 2.19). Physical modeling also exploits this new instrumentation.

The founding of the NSF-supported CSDMS (Section 2.3) will encourage the development of sediment transport law theory and its incorporation into landscape evolution models. New models exploiting parallel processing and supercomputers may permit retention of fine-scale process laws over applications on large spatial and temporal scales. The

FIGURE 2.18 Does topography cause bedrock to break? Processes on bedrock-dominated landscapes such as the one pictured here are as yet poorly understood. There are intriguing hints that rock strength, topographic form, erosion rate, and debris production may be linked and therefore coevolve. SOURCE: Photograph courtesy of William E. Dietrich, University of California, Berkeley.

NSF funding of three Critical Zone Observatories (CZOs; Box 2.5), which are dedicated in part to observing and modeling the evolution of soil and weathered bedrock, will advance significantly our understanding of weathering and landscape evolution. Furthermore, advances in various geophysical tools are enabling us to monitor from space and explore the subsurface in entirely new ways (Box 2.6).

Rate Laws Explicitly Accounting for Geology, Climate, Biota, and Land Use

A challenge to all geomorphic law formulations is how to account explicitly for geology (bedrock type), climate, biota, and land use. This list is surprising because

FIGURE 2.19 *Left:* Scientists working in Hawaii are employing repeat lidar surveys and a suite of sensors to investigate how land-cover changes in tropical watersheds affect coral ecosystems and coastal habitats. *Right:* Difference map showing erosion amounts between subsequent lidar scans. Lidar surveys combined with isotopic analyses and data from overland flow sensors, rain gauges, soil moisture probes, and suspended sediment sensors were used to calibrate a geomorphic transport law for erosion by overland flow. SOURCE: Images courtesy of Jonathan Stock, U.S. Geological Survey.

qualitatively, even introductory textbooks recognize specific roles that rock properties, climate conditions, and biotic activity play in Earth surface processes and landscape evolution. We lack, however, metrics that quantitatively link landscape form to these controls. Furthermore, few studies define how specific aspects of bedrock properties, climate, or biota should be included in transport laws. A small number of pioneering models have been proposed that include these major controls. In all parts of this chapter, the development of transport laws is important, especially transport and erosion expressions that apply at shorter time scales. Although not highlighted further here, the role of bedrock in influencing hydrologic, ecologic, and geomorphic processes has not yet been put on a mechanistic footing. Many ecologic and geomorphic processes may require formulations based on short time scales (such as storm events or actions of individual organisms) that then can be applied over long time scales (for landscape evolution modeling) or to land-use analysis. These kinds of processes point to fundamental questions about what characteristics of climate, bedrock, or biota strongly influence processes and how these characteristics can be incorporated into transport laws and modeling. The processes by which bedrock, climate, and biota dictate landforms remain frontier problems in Earth surface processes.

BOX 2.6
The Importance of Shallow Geophysics

Understanding processes that influence the chemistry, form, and function of landscapes requires quantification of rates of mass transfer and the composition and structure of the Earth's surface. For example, quantifying the sediment flux down a hillslope and its relation to spatial and temporal variations in vegetation and precipitation is important for improving a hillslope geomorphic transport law and for geochemical models of regolith formation. Equally important is the subsurface structure of landscapes, including the depth to unweathered bedrock, fracture density of the bedrock, and spatial variations in the composition of weathered and unweathered hillslope material. Numerous remote-sensing and ground-based geophysical methods have been developed to quantify these aspects of the near- and subsurface environments (see figure below). *Hydrogeophysics*, for example, has recently emerged to develop tools to improve subsurface monitoring of bedrock moisture dynamics (Section 2.6). The focus of many shallow surface geophysical studies is on environmental engineering problems, such as contaminant transport in shallow aquifers or satellite-based observations of ground cover. As described in a 2006 CUAHSI review of geophysical instrumentation for watersheds, advancing shallow geophysics tools for the near-surface bedrock environment and making them widely available for research investigations is important (Robinson et al., 2008). The content of this report highlights new and exciting research opportunities for the shallow and remote-sensing geophysics communities to engage in the study of Earth surface processes. The recent formation of a new focus group, Near Surface Geophysics, in the American Geophysical Union attests to some of the new interest in this type of research. Extending these techniques to enable measurement of geochemical and biological properties of natural systems will become increasingly important.

A diverse range of ground and airborne geophysical techniques are applicable to study the evolution of the form, composition, and function of landscapes. Tools commonly used in the environmental engineering communities include ground penetrating radar, high-frequency seismic methods, microgravity, and a range of electrical and magnetic methods (see figure below; also Appendix C). These tools can image variations in subsurface physical properties (such as density, electrical resistivity, and acoustic velocity), all of which relate to variations in soil, ice, or bedrock composition and/or the presence of air or water in pores and fractures. A wide range of ground or airborne remote-sensing techniques is also available, and new techniques are under development. Remote-sensing techniques have proven useful for imaging numerous characteristics relevant to Earth surface processes. Examples include high-resolution imaging of topography, topographic and ice surface change over time, and vegetation; imaging surface velocities of glaciers and mass movements; multispectral imaging of Earth's entire surface and atmosphere every one to two days; and measurement of spatial and temporal variations in rainfall. Focusing these and other new and developing geophysical techniques toward studying the evolution and operation of Earth's surface is important.

Interdisciplinary collaboration between Earth surface scientists, geophysicists, and engineers is needed to advance these emerging technologies, provide training, and make such geophysical instruments widely available. The availability of global positioning systems and seismic instruments through the University Navigation Satellite Timing and Ranging (NAVSTAR) Consortium (UNAVCO) and Incorporated Research Institutions for Seismology (IRIS) is one model of what could be done for shallow geophysics. Examples of exciting interdisciplinary research problems in Earth surface processes that can be addressed through geophysical studies include (1) quantifying variations in soil and weathered bedrock thickness and the physical, biological, and chemical processes controlling these variations; (2) quantifying subsurface moisture and flow dynamics that control physical and chemical erosion; (3) documenting fracture density in bedrock, quantifying its influences on hydrology, weathering processes, and denudation; (4) near-real-time monitoring of mass transport across landscapes to improve geomorphic transport laws and geochemical models; (5) quantifying temporal and latitudinal variations in the rates and magnitudes of periglacial physical properties and processes in the presence of global warming; and (6) quantifying temporal changes in ice surface elevation and velocity, sub-ice sheet hydrology, and calving processes at the ocean-ice sheet interface.

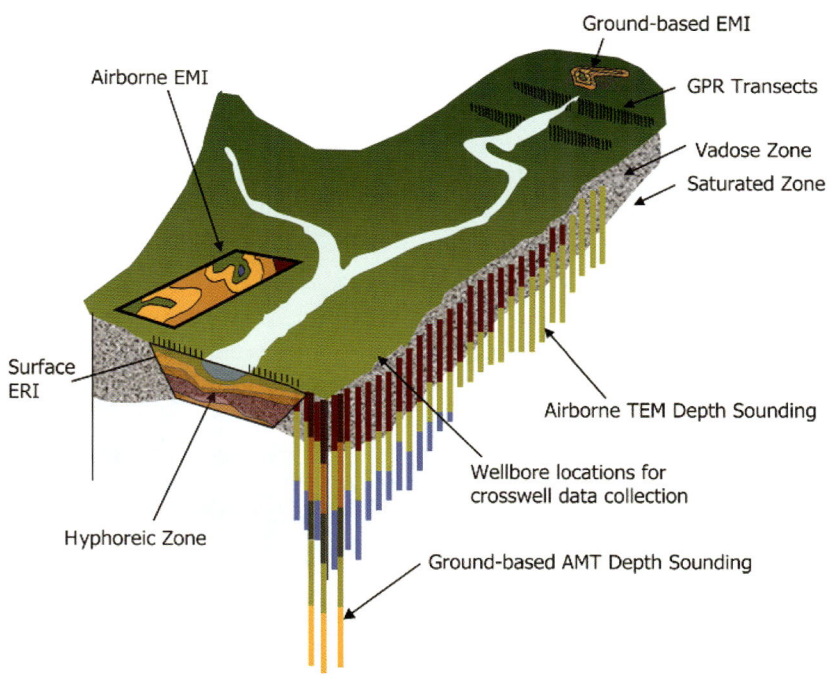

Several airborne and ground-based electromagnetic geophysical imaging techniques can be used to characterize subsurface materials and hydrologic dynamics, including ground penetrating radar, electrical resistivity imaging, transient electromagnetic, and audiomagnetotelluric techniques. The combination of these methods can build observations from local sites to entire watersheds. SOURCE: Robinson et al. (2008); ©2008 D. Robinson. Reproduced with permission of John Wiley & Sons, Ltd.

2.6 HOW DO ECOSYSTEMS AND LANDSCAPES COEVOLVE?

How different would Earth be if life had never evolved? Will the biotic response to future climate change drive significant changes in landscape processes? Biota break rocks into soil particles, exhaust gases into the atmosphere, and alter runoff chemistry (see also Section 2.4). The movement, growth, addition of organic matter, and dissolution associated with biota make otherwise dense or perennially frozen ground permeable and cause rainfall and snowmelt to enter into the subsurface. Flow of water into the subsurface may reduce flood peaks and later sustain low flows during dry periods. Organisms also weave strength into weak materials, stabilizing hillslopes against erosion, and confining river flow within narrower channels. Life—through digesting, dilating, exhaling, decaying, pushing, and weaving—strongly influences the form and pace of surface erosion and strongly modulates biogeochemical cycling. Hence, life simultaneously affects climate, hydrology, erosion, and topography (Figure 2.20; see also Sections 2.2, 2.3, and 2.4).

Despite our awareness of the role of life in Earth surface processes, we often have only a qualitative appreciation of this connection. As discussed elsewhere in this chapter, however, biota are important and, in several cases, central. Questions such as those posed above call for a mechanistic understanding that can form the foundation for making predictions or performing modeling experiments to explore and discover the richness and importance of these interactions. The incorporation of the biosphere into global climate models, for example, transformed these models, enabling them to address accurately such vital issues as the controls on CO_2 buildup. The climate community accomplished this task by model building, monitoring, and large-scale field campaigns to obtain data on controlling mecha-

FIGURE 2.20 Do trees influence the shape and surface composition of mountains? SOURCE: Courtesy William E. Dietrich, University of California, Berkeley.

nisms. A similar effort is needed in the Earth surface processes community in conjunction with the development of some of the research opportunities outlined below.

The quantitative study of the interactions of biotic, physical, and chemical processes in the Earth's surface, or *geobiology*, is in its infancy. The most active area of research into the connections of biotic processes with inorganic Earth materials is at the microbial level, and research has focused on discovering the composition and functions of these organisms and their influence on the environment. This effort is a central part of the emerging field of *geomicrobiology*, and many Earth sciences departments have made recent hires in this area. Microorganisms were the only life form on Earth for most its history, and the influence of microorganisms on the Earth's geochemical cycles remains profound (see also Section 2.4). Since the earliest stages of Earth's evolution, methane gases produced by bacteria may have significantly elevated the atmospheric temperature, keeping it warm even when the then-young Sun was 30 percent less luminous. The evolution of some forms of photosynthesis by microorganisms drove up oxygen levels, enabling a vast diversity of life forms to emerge. Interest in the influence of microbes on geochemical cycles has increased with the discovery of water on Mars, the methane lakes of Titan, and numerous extrasolar planets—all of which raise the question of the possibility of life elsewhere. Although research in geobiology has been expanding rapidly, little work has been directed toward the effects of microbial activity on the form, composition, and evolution of the Earth's surface. Some studies have shown that the surface crust formed by microorganisms in arid landscapes may strongly influence runoff and erosion processes. At present, however, we do not know if microbes affect the shapes of hillslopes or the form of rivers, nor can we quantify how biotic activity influences ecosystem services.

We do know that plants and animals such as trees, worms, gophers, and insects strongly influence surface processes and consequently the form and composition of the surface and ecosystem functioning. In many forested environments, trees root into the underlying bedrock, and if the tree falls, the roots tear the bedrock, converting this rock into soil material, transporting both soil and rock downslope. Through such processes, forests make and move their own soil. Charles Darwin devoted his last book to the action of worms in which he quantified the rate of formation and transport of soil by worms and concluded, "When we behold a wide, turf-covered expanse, we should remember that its smoothness, on which so much of its beauty depends, is mainly due to all the inequalities having been slowly leveled by worms."[4] While soil formation rates and soil transport rates by biota have been quantified in some settings, we lack theory to guide prediction of these process rates and have limited ability to include biota mechanistically in landscape evolution models.

Strong coupling of life and landscapes also occurs in river systems and marshlands.

---

[4] C. Darwin, 1881. *The Formation of Vegetable Mould Through the Action of Worms with Observation of Their Habits*. London: John Murray, 328 pp.

Recently several research groups have presented field and laboratory data demonstrating that riparian vegetation, while seeking the moisture, nutrients, and light along river valleys, typically adds considerable strength to banks, traps and slows overbank flows, drops clumps of roots and trunks into the river that armor the bank, or can float away and become trapped into jams of debris. Through these actions, vegetation may transform wide channels with multiple bars and flow pathways into single-thread channels or convert the channel into multiple vegetation-lined threads (see also Section 2.2). The vegetation can control the rate of channel shifting and, ultimately, direct the formation of floodplains. So significant is large woody debris in channels to creating habitat that millions of dollars are spent annually adding wood to streams to recover lost ecosystem functions due to past wood removal (see also Section 2.8). In marshlands, vegetation plays a primary role in controlling fluxes, deposition patterns, and channel development. An understanding of this coupling is crucial to predicting the fate of tidal marshlands during the sea-level rise expected with climate warming and designing successful marshland restoration projects (see also Section 2.9). Great theoretical and practical questions now present themselves with respect to how riverine or marshland ecosystems and their channel morphology and dynamics are coupled.

The study of the coupling of ecosystem and hydrologic processes has recently been labeled *ecohydrology* and is rapidly expanding as a field of research. This coupling is dramatically played out in the water-limited environments of arid landscapes. Arid zone vegetation is often found to be patchy, and various models to explain this have been proposed that argue for codevelopment of vegetation patches, runoff and infiltration, soil moisture, and erosion. Closely coupled to the presence of vegetation, fluxes of water in arid regions also directly influence salt contents of soils, which can control soil permeability. Connections to human activity are also involved. Some have proposed that land use has reduced the crust and vegetation cover of arid lands, leading to more dust erosion, and some of that sediment is deposited in mountain glaciers (as in the Rocky Mountains) leading to lower albedo and greater melting. Only a few recent papers have argued for the use of the word *ecogeomorphology*, but the idea of expanding these studies to include landscape evolution is established and research is expanding rapidly. One example of such research is the exploration of the coupled ecologic, hydrologic, and geomorphic evolution of hillslopes in landscapes where biologic communities are strongly influenced by hillslope orientation (aspect) and hillslopes display a distinct topographic asymmetry. Such research at the interface of hydrology, geomorphology, ecology, and geochemistry is at its inception, but shows great promise.

The coupling of life and landscapes may seem least direct when considering entire ecosystems but, in fact, may be most crucial to understanding the large-scale evolution of topography. As land masses collide and elevate rocks, life may be more than a passive passenger as mountains build and are eroded. Biotic processes directly affect the form and rate of erosion. Bedrock breakdown, soil development, and hillslope erosion—all driven or

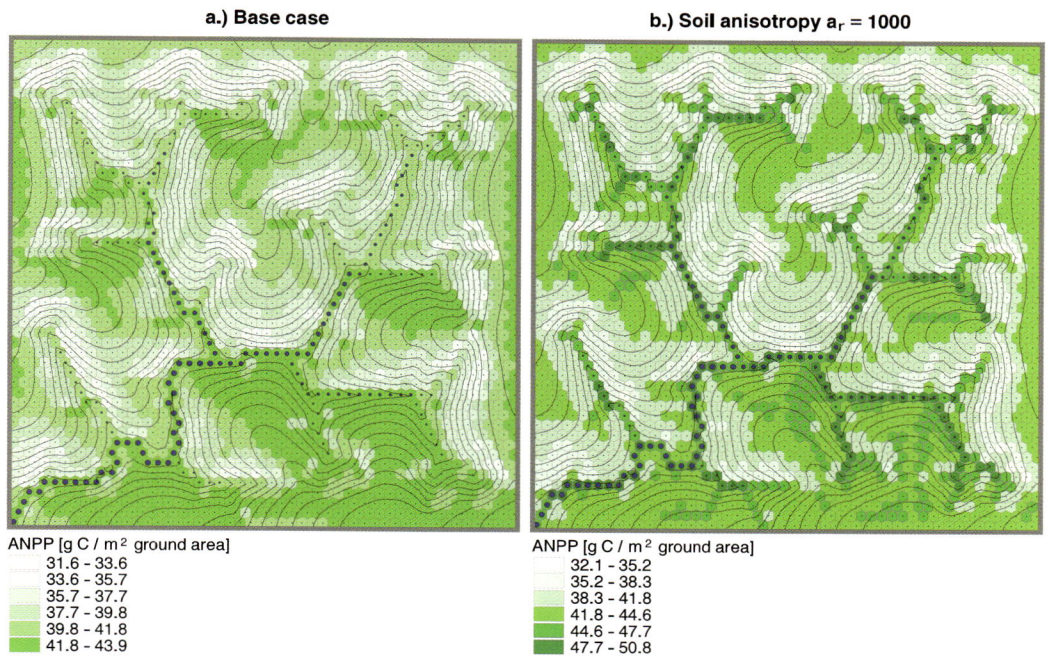

FIGURE 2.21 Model results showing annual net primary productivity of grass predicted on a landscape evolved from a specific set of erosion rules. SOURCE: Ivanov et al. (2008); reproduced with permission of the American Geophysical Union.

mediated by biota—strongly influence the size and composition of sediment entering rivers (see also Section 2.4). Sediment load and particle size impose constraints on river slopes, and steeper channels will tend to lead to higher mountains. Vegetation extracts moisture from the ground and exhausts it into the atmosphere, cooling the air, and nourishing clouds that later return the moisture to the ground as rain and snow. This exchange process affects the global heat balance—hence atmospheric circulation and the spatial pattern of precipitation. This pattern in turn may direct both chemical and physical erosion rates and, on mountains, affect the spatial pattern of unloading. The pattern of erosional unloading directs the evolution of mountain systems, and by these connections, biota may influence the shape, height, symmetry, and surface chemical composition of mountain ranges. To explore quantitatively the evolution of landscapes and climate (see also Section 2.3), the effects of biota need to be specifically and mechanistically included in models (Figure 2.21).

Recent developments indicate that we now are ready to make significant advances in understanding the coevolution of landscapes, life, and ecosystems. In addition to the emergence of the fields of geobiology, ecohydrology, and ecogeomorphology, new efforts are under way in field observatories that explicitly address linkages between biota, Earth surface

BOX 2.7
National Center for Earth-surface Dynamics

NCED is an NSF-funded Science and Technology Center (STC) created to catalyze development of an integrated, predictive science to examine processes that shape the Earth's surface and, specifically, to predict the coupled dynamics and evolution of landscapes and their ecosystems (http://www.nced.umn.edu/). NCED research is organized into three integrated projects: (1) *Desktop watersheds* seeks to exploit the spatial structure imposed by tributary channel networks, expressed by high-resolution topography, to provide static and dynamic predictions of local physical, geochemical, and ecosystem properties. (2) *Subsurface architecture* uses information from modern systems, experiments, and stratigraphic records to develop a predictive understanding of delta evolution and apply this understanding to delta restoration. The work will also improve prediction of variations in porosity and permeability that control the flow and accumulation of water, oil, and gas in the subsurface. (3) *Stream restoration* addresses the scientific basis for this multibillion-dollar activity in the United States through a combination of research and training developed in coordination with agency, industry, and academic partners including the U.S. Geological Survey, U.S. Army Corps of Engineers, Environmental Protection Agency, U.S. Department of Agriculture, U.S. Bureau of Reclamation, and Bureau of Land Management. The goal is to move restoration practice away from reliance on analogy to an analytical, process-based approach.

Knowledge transfer, education, and *diversity* are all integrated into NCED's research programs. Knowledge transfer includes exchange and engagement with the broader research community via workshops, working groups, a visitor's program, short courses, and postdoctoral researchers. NCED's *education* program uses the familiarity and aesthetic appeal of landscapes to engage a broad spectrum of learners in NCED science. The centerpiece of the NCED education program has been collaboration between the Science Museum of Minnesota (SMM) and NCED's academic institutions and applied partners. One outcome has been the successful *EarthScapes* exhibit (http://www.nced.umn.edu/Earthscapes.html) at SMM. A major new traveling exhibit initiative, *H2O: Water = Life*, developed by SMM and the American Museum of Natural History in New York, along with three STCs, is also helping the public learn about Earth surface processes (http://www.amnh.org/exhibitions/water/). The NCED-SMM collaboration led to the NSF-funded *Future Earth Initiative* (http://www.smm.org/exhibitservices/history/futureEarth/), which will serve as a center for informal education activities on human influence on the environment. NCED's *diversity* program addresses the mismatch between the current spectrum of participants in environmental science and the U.S. population overall. Working with Ojibwe tribal elders, NCED has developed environmental camps that use innovative, culturally sensitive programming to excite Ojibwe children about environmental sciences and encourage them to excel in school and pursue science-related careers. NCED also has a vigorous recruiting program for minority participants in its research program, and minority participation increased from 8 percent when the program started to 17 percent in 2008.

processes, and landforms. The National Center for Earth-surface Dynamics (NCED) was established in 2002 to advance a predictive Earth surface dynamics science and specifically seeks to apply this predictive capability to ecosystem and land-use management (Box 2.7). Three CZOs were initiated by NSF (see Box 2.5) to quantify and predict the interactive processes occurring in the zone between the vegetation canopy and the underlying groundwater table (with emphasis on chemical weathering, biogeochemical cycling, soil formation, and hydrology), and the CZO network is growing to provide field sites for study along environmental gradients. Explanation of weathering and soil formation patterns, biogeochemical cycling, vegetation-mediated hydrologic dynamics, and landscape evolution at these observatories will require collaborations across many disciplines. Twenty-six Long Term Ecological Research (LTER) network sites also conduct ecological research that integrates ecologic with geologic, hydrologic, and atmospheric sciences (see also Section 2.4 and Chapter 4). Research questions at several LTERs specifically focus on the biological-physical feedbacks that shape landscapes at decadal or longer time scales. The proposed National Ecological Observatory Network (NEON) has identified 20 potential sites (about 225 km^2 each) across the United States to monitor ecosystems and their drivers for the next 30 years. In addition, as many as 60 shorter-term monitoring sites may be established. Although the coupling to Earth surface processes and landscape form, composition, and evolution is not explicitly a goal of NEON, the data gathered will provide a wealth of information for such endeavors. An important opportunity for critical interdisciplinary advances may be missed, however, if biologists focus on NEON sites, Earth scientists focus on CZOs, and hydrologists develop their own observatories. In May 2008, the Meeting of Young Researchers in Earth Sciences (MYRES) held an international conference on the "Dynamic Interactions of Life and Its Landscape" attended by more than 60 participants (see also Preface). These current graduate students and recent doctoral degree recipients reported on laboratory, field, and numerical modeling studies that are beginning to lay the foundation for this emerging field.

2.7 WHAT CONTROLS LANDSCAPE RESILIENCE TO CHANGE?

Geological and modern records of sediments and landforms indicate that Earth's surface changes under the influence of external drivers such as climate change, tectonics, volcanism, and human activities, as well as the internal dynamics of erosion and deposition acting over long periods of time (Figure 2.22). Some changes are gradual; others are abrupt or rapid, ranging over time scales from the seconds in which a landslide occurs to the millennia over which glaciers move across a mountain range and dramatically change the landscape. As used in this report, "abrupt" change refers to a response that is nonlinear and much faster than the forcing, while "rapid" change refers to a response that can be linearly related to forcing.

FIGURE 2.22 The Aletsch glacier in Switzerland is the largest glacier in Europe and was recently designated as a UNESCO World Heritage Site. This glacier and its predecessors have dramatically altered the landscape through repeated glacial-interglacial cycles over the last ~2 million years. The pulsing of glaciers across landscapes represents an abrupt change in state and a switch in the dominant erosive geomorphic agent. SOURCE: Courtesy of Todd Ehlers, University of Tübingen, Germany.

The shape of landforms and their rates of evolution fluctuate within ranges reflecting the stochastic nature of the processes and materials that drive their operation. A single-thread meandering river channel will fluctuate in width, depth, or sinuosity as it is perturbed by floods, by fluctuations in sediment supply, and by its own bend-lengthening and bend-cutoff processes. Channel cutoffs may occur and form lakes in a floodplain, yet the meandering channel can retain a remarkably similar geometry for centuries, even as the channel migrates across its floodplain.

Under other circumstances, changes in external drivers and internal dynamics of landforms intersect to impose a new set of processes on the landscape. If riparian and floodplain vegetation is thinned by drought, the floodplain may become erodible by overbank flows, the bank strength may be reduced, and a multithread, less sinuous channel may develop with frequent avulsions, a steeper gradient, higher sediment transport rate, and a higher rate of channel and floodplain change. Such a transition is often labeled "channel instability," but it represents an enduring change in the probability distributions of channel forms and operations, under an altered set of processes, driven ultimately by a change in climate and vegetation in this example.

Landscapes generally fluctuate within some predictable range in response to variation in external drivers and interaction among internal processes as long as the processes that alter it remain below some level of intensity. When conditions change with sufficient magnitude and duration, landscapes may become altered beyond the range within which they can recover. The driving agents have then exceeded the landscape's resilience to change, resulting in change to another form with a different intensity of erosion, sedimentation, or dominant morphogenetic process. Through such changes, the forms of landscapes can themselves experience sudden changes, such as from extensive and smooth, aggradational features (alluvial fans and valley fills) to sharp-edged, incising forms (steep escarpments with landslide scars, dense valley networks, or gullies) (Box 2.8). The material-transporting systems of Earth's surface can also be intensified or even reorganized so that, for example, a river-dominated coastline with a copious sediment supply can have its sediment regime transformed into one of net sediment loss when a delta lobe switches position or river sediment is impounded by dams. The resilience of surface features to change may be defined (in either observed or conceptual terms) by the magnitude of the disturbance that they can absorb before changing form or behavior.

Progress has been made in the past two decades in developing numerical models that simulate erosion and sediment transport processes, driven by climate and land-use changes. The models predict spatial and temporal patterns of landform evolution, sediment supply to river channel networks, and the conditions required for switching between periods of sediment accumulation and removal. More recently, emphasis has been placed on applying these models to real river basins in regions for which it is possible to provide time control on rates of erosion and sequences of sedimentation. The models are driven by long-term weather records. Some of the models simulate how vegetation cover is also driven by climate changes. The model results can then be compared to sedimentary records of climate change to understand, for example, the range and magnitude of conditions that a landscape has experienced over time.

This section highlights the state of scientific knowledge related to rapid and abrupt change on Earth's surface and why some landscapes undergo a "change of state" whereby their form may change dramatically or different morphogenetic processes may become dominant. The section also highlights the needs and opportunities for investigating controls on the resilience of landscapes to change (see also Section 2.9).

Some areas of Earth's surface are more vulnerable than others to changes in state. Polar, glacial, and periglacial regions are currently nearing, or are in, a change of state and are predicted to continue to do so with persistent global warming. Paleoclimate records show temperature swings between glacial and interglacial periods that are several times stronger in the polar regions than in the tropics because of greater sensitivity to orbital change and snow and ice feedbacks. Changes in the state of the landscape also pertain to areas in which concentrations of human population impose increasing pressures on Earth's surface environments (see also Section 2.8). The goals and challenges of research in Earth surface processes include identifying thresholds of change, understanding the environmental processes most

BOX 2.8
The Future of Landslide Prediction

The future of landslide hazard research lies in making predictions about when, where, and how big the landslides will be, and in developing procedures that convey warnings and risk to the public so that loss of life and damage to infrastructure or ecosystems are reduced. Such effort is critical, and particularly important in anticipating the effects of climate change and land use in areas prone to slope instability (see figures below). Land use, especially road construction and vegetation clearing (including industrial timber harvesting), has led to great increases in landslide frequency. The underlying causes for this are well understood and landslide theory and empiricism can be used to guide management decisions.

Landslides occur in sizes that vary from a sand pile avalanche of a few centimeters to entire landscapes involving many cubic kilometers of earth material. They can move slowly, creeping along in wet years. They can remain stable for decades or millennia, then collapse suddenly and travel at meters per second. Entirely new landslides may occur where no previous evidence of landsliding exits. The risk and management response differs with scale and speed. This variability of scale and speed, while challenging to anticipate, is not random. Types of landslides generally take on certain patterns that can be described statistically (see Section 2.2).

At present much can be done to delineate where landslide hazards are greatest by mapping historical occurrences of landslides and quantifying correlations between occurrence and probable controlling factors (e.g., hillslope gradient, geology, and vegetation). These correlations can be applied across large areas to identify where landslides are likely to occur in the future. Such correlations have improved in the past 20 years with the availability of digital spatial information for attributes such as topography, geology, vegetation, rainfall, soil depth, and other factors. High-resolution topographic data obtained from airborne laser mapping greatly increase our ability to identify and delineate landslides and to predict their future occurrence. Hazard maps based on correlations may have limited ability to predict future events under changing climate and land use, hence fundamental research on landslide mechanics remains essential.

Landslides commonly are caused by elevated water content in the ground, which reduces the strength of materials. Landslides also are caused by ground shaking during earthquakes, and can contribute significantly to the loss of life and destruction of infrastructure. Rainstorms, rapid snow melt, ground ice melting, and simply wetter years can lead to elevated water content and slope destabilization. Various international research groups are exploring the use of precipitation forecasts (from global or regional climate models) to predict when an area may experience landslides. These forecasts are often developed at rather coarse scales and require methods to predict local precipitation. Once the forecast is made, it needs to be translated into landslide likelihood. The simplest approach has been to use empirical relationships, typically relying on analyses of the rainfall intensity and duration at which landslides occur. These predictions can be made more specific by using landslide hazard maps to identify more sensitive parts of the landscape where such landslides are most likely to occur. Real-time monitoring of precipitation, water content, and even ground deformation is being done in some areas to add further information. Advances in fine scale (in space and time) precipitation forecasts and real-time monitoring, coupled with landslide potential mapping, have led to landslide warning systems being tested in various countries.

A major challenge remains in predicting the specific size and location of landslides for a given precipitation event. All empirical and theoretical procedures tend to over-predict the area that may fail in a given rainstorm. Poor knowledge of the spatial pattern of subsurface conditions that control failure also limits our predictive capabilities. Basic research is still needed in this area.

The emergence of new technologies (lidar for topography, radar for rainfall, real-time monitoring, and advanced computing) facilitates empirical forecasts of when and where (in a regional sense) landslides might occur. Future advances will allow us to refine predictions of when, where, and how big landslides might be. Models that treat not just the failure of earth materials but also the deformation and movement from initiation to cessation will increase our understanding of mechanisms. These many efforts will spur progress towards the goal of reliable hazards maps and warning systems.

Landslide destruction of La Conchita, Ventura County, California. *Left*: In 1995 two successive landslides over a one-week period carried about 1.3 million m^3 (1.7 million yards^3) downslope, damaging or destroying 14 houses. *Right*: In January 10, 2005, on a sunny day after a two-week period of heavy rain, a landslide struck the community again, destroying or seriously damaging 36 houses and killing 10 people. About 200,000 m^3 (250,000 yards^3) of remobilized debris rushed through the community at a speed approaching 30 feet/second (Jibson, 2005). SOURCE: U.S. Geological Survey.

vulnerable to change, understanding the mechanisms that make some landscapes resilient to change, and anticipating and investigating options for mitigating or even reversing the effects of change.

Studying the impact of abrupt changes on Earth's surface requires looking into geologic records of their occurrence. As discussed in Section 2.1, studies of Earth surface response to past changes foreshadow potential future change and clarify the evolutionary trends that produced today's landscapes. Documenting the occurrence and impact of rapid or abrupt climate change events on landscapes is particularly challenging in that it requires high-resolution paleoclimate reconstructions to complement stratigraphic records, as well as documentation of Earth surface response encoded in the physical, chemical, and biotic properties of current landscapes. While new high-resolution marine and ice-core records are continually being collected for paleoclimate research and other purposes (Box 2.9), less work has been done on terrestrial (paleosol) records, which can be more sensitive to geographical and latitudinal variations in climate.

In the coming decades, significant advances will be required in the measurement of past and present rapid landscape change (via field and remote-sensing studies) and in improved predictive models. The tightly coupled response of landscapes to changes in climate, tectonics, biota, and human activity at multiple spatial and temporal scales requires increased emphasis on interdisciplinary research to take advantage of emerging opportunities and tackle the challenges highlighted below.

Disappearing Glaciers and Glaciated Landscapes

Over the past several decades, mountain glaciers and continental ice sheets have retreated rapidly. Accelerated melting of glaciers has contributed to sea-level rise, increased numbers of glacial lakes, glacial-flood outbursts, altered river flows, and changed flooding patterns. Over millennial to million-year time scales, glaciers are powerful geomorphic agents as glacially carved and glacially deposited landscapes attest. Several challenges, highlighted below, face the scientific communities studying ice sheets (for example, Greenland and Antarctica) and mountain glaciers.

Current models of continental ice-sheet dynamics match overall ice-sheet trends observed during the twentieth century, but do not predict the observed rapid marginal thinning currently taking place. Such a deficiency implies that understanding of the mechanisms controlling ice dynamics, particularly with regard to mechanisms that accelerate ice flow as temperature approaches the threshold for melting, is inadequate. Current models could, therefore, be inaccurate in predicting future loss of ice sheets and long-term rates of glacial erosion. This problem is a leading cause of the uncertainties in projected future global sea-level rise; the supply of buoyant freshwater to the North Atlantic; and the resulting risk of abrupt climate change due to altered ocean circulation patterns. The deficiency of

BOX 2.9
Paleoclimate Records of Abrupt Climate Change

Abrupt climate change is more frequent in the geologic record than previously thought. Because precipitation carries with it the isotopic signature of the air from which it fell, snowfall trapped in ice cores over thousands of years progressively records, among other chemical factors, the oxygen isotope variations in the atmosphere relative to the standard ocean water oxygen isotopic ratios at the time the snow fell. These ratios can, in turn, be correlated to variables such as temperature at the time of the snowfall. Long-term ice core records can thus show seasonal to long-term variations in temperature and other conditions. The figure below shows the $\delta^{18}O$ ("delta oxygen 18" or the change in ^{18}O concentration) versus depth and age in the Greenland Ice Sheet Project Two (GISP2) ice core. The large variability in $\delta^{18}O$ shown in this record is due to a number of factors, the most significant of which are air temperature and changes in ice volume. Time series analysis of this record has identified 23 rapid fluctuations that occur every ~1,470 years (±532 years) between 110,000 and 23,000 years before present (B.P.). These events are called Dansgaard-Oeschger events during the last glacial period and Bond events over the last ~10,000 years (Holocene). They involve decade-long warming events followed by gradual cooling over hundreds of years.

The spatial extent and potential causes of these climate fluctuations beyond the Greenland ice sheet are active areas of research because of the impact they could have on landscapes and people. For example, the Little Ice Age that occurred 200 to 400 years B.P., with well-documented damaging impacts on European society, has been interpreted as part of the cold stage of a Dansgaard-Oeschger event. Abrupt droughts have also been identified in the tropics.

Documenting past occurrences and triggers of such abrupt climate changes is currently a high-priority topic of study for the paleoclimatology and glaciology communities (Cox, 2005). Abrupt climate changes should have a large and measurable impact on the form and evolution of Earth's surface, but the capacity for predicting their occurrence and probable consequences is limited. Studying the impact of abrupt climate change on landscapes is an interdisciplinary effort and is closely coupled to paleoclimate studies.

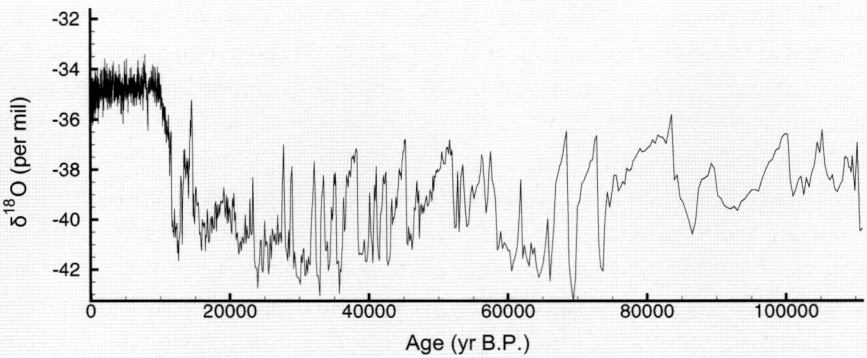

The $\delta^{18}O$ profile of the GISP2 ice core from the summit of the Greenland ice sheet. SOURCE: Data from Grootes and Stuiver (1997); reproduced with permission of the American Geophysical Union.

the ice dynamics models is attributed to inadequate understanding, and thus lack of treatment, of subglacial hydrology, basal sliding, subglacial sediment deformation, and ocean-ice interactions at ice-sheet margins (Iverson, 2008). Future efforts to improve coupled ocean-atmosphere-ice sheet models will incorporate the processes summarized above. Parallel efforts in satellite-, ground-, and marine-based studies are also needed to provide data for model evaluation. Additional quantification of chemical and physical processes in periglacial environments could also improve predictability of changing chemical fluxes of carbon and methane as well as physical changes that have impacts on societal infrastructure.

More robust predictions are also needed to evaluate mountain glacier contributions to sea-level rise and how climate change will influence mountain glacier interactions with the hydrologic cycle and alpine natural hazards. Central to these scientific challenges is a need for an improved quantitative understanding of the long-term evolution of alpine landscapes (Owen et al., 2008). The main tasks confronting the community include quantifying processes controlling glacier motion and erosion (such as glacial hydrology and sliding); clarifying past spatial-temporal patterns of mountain glacier fluctuation in relationship to regional climate and ocean-atmospheric circulation; continued development and testing of numerical mountain glacier models and their effects on mountain landscapes over repeated glacial-interglacial cycles; characterizing the role of climate variability on glacier variations; and predicting the distribution, size, and nature of glaciers in the future.

Thawing Permafrost

At high latitudes and in high-altitude mountainous environments the regolith and underlying bedrock are permanently frozen except for a seasonally thawed "active layer" near the surface. Global warming is rapidly changing permafrost environments and has contributed to an increase in the active layer thickness resulting in rapid changes in periglacial landscapes and landforms. Permafrost thawing has changed patterns of soil organic carbon and nitrogen content, vegetation, biogeochemical processes, rates of permafrost creep, and mass movements. The impacts of these changes on the form, function, and evolution of periglacial environments and the atmosphere are not well known.

Frozen and unfrozen water is disseminated throughout the regolith, while in other parts ice is highly segregated into layers and "frost wedges," several meters thick. The low temperatures and evaporation rates at these latitudes maintain the regolith in a moist condition, so that some of the carbon fixed by photosynthesis each year accumulates in soil. The amount of this accumulation has varied since deglaciation, and a considerable amount of field research and modeling has recently been invested in reconstructing the postglacial history of accumulation of this vast carbon reservoir, which currently stores about 14 percent of soil carbon.

The carbon reservoir emits both methane and CO_2 to the atmosphere, directly through the soil surface and by concentrated bubbling through lakes. The future stability of this carbon reservoir in the presence of global warming and whether or not the carbon reservoir will be vented as methane or as CO_2 are major uncertainties in climate projections (see Section 2.4). Thus, modeling the history and future of soil carbon storage and linking these permafrost models to climate models is important research and requires more information about the distribution and age of permafrost and its relationship to groundwater and lake hydrology across northern regions. Concern about the future of the permafrost and its carbon reservoir has been raised by the recognition that permafrost in interior Alaska has warmed by approximately 1.5°C since the 1980s and that temperatures in boreholes on the North Slope of Alaska rose by 2-4°C during the past century. In addition to being driven directly by the atmospheric radiation budget, permafrost is also thawing around lakes as they warm and along coasts as a result of wave erosion of frozen but unconsolidated sediments as sea level rises and as marginal sea ice melts and no longer buffers wave action.

The landscapes of the northern permafrost regions have largely been created by sediments left behind from glacial and fluvioglacial processes. In areas where the permafrost is continuous or at least widespread, warming is leading to greater seasonal thawing and drainage of groundwater, thereby causing the creation or expansion of lakes as the regolith collapses upon melting. In areas of discontinuous and sporadic permafrost, remote-sensing studies have revealed a decrease in lake area due to the thawing of their substrates and breaching of new outlets. The fate of lakes in subarctic regions has an important influence on the evasion of methane and thus potentially on atmospheric chemistry and the radiation budget. Quantifying the chemical and physical processes active in permafrost environments relies on continued development of physically based, process-oriented models as well as mapping and characterization of permafrost depth and distribution from multiple geophysical, geochemical, and field-based techniques.

Coastlines in Transition

Coastal areas are the dynamic interface between land and ocean. Coastal areas worldwide are experiencing eustatic sea-level rise that is exacerbated in many areas by subsidence of already low-lying land surfaces. The current situation is alarming because of the number of people living in lowlands within 150 kilometers of the coast; this number has been projected to increase from 3.6 billion in 1995 to 6.4 billion, or 75 percent of the world's population, by 2025.[5]

Existing estimates of land-sea elevation changes have been made without considering how the effects of subsidence of coastal land areas, altered sediment transport and deposition, or changes in the biosphere may interact with eustatic sea-level rise, which itself varies

---

[5] http://www.aaas.org/international/ehn/fisheries/hinrichs.htm.

around the globe because of the combination of isostatic, tidal, and rotational effects. The combined effects of these processes, in addition to climate change, may amount to several meters of relative land-sea level change over the coming century.

Many coastal zones are especially sensitive to sea-level rise because, with low relief, small vertical changes in sea level affect extensive areas. As sea level rises, low-lying areas are subject to tidal inundation, flooding by storm surge, and increased wave energy due to greater water depth. These processes alter the relationships between sediment deposition and erosion by altering the productivity, distribution, density, and types of vegetation and animals, as well as the physics of the coastal system. Tidal marshes are obvious examples of coastal landscapes in which ecological and physical processes are tightly coupled. Marsh vegetation (type, density, productivity) depends on the relationship between surface elevation and tidal range, while marsh vegetation influences water velocity, sedimentation, and erosion of the marsh surface. How large do changes in hydroperiod or wave energy need to be to alter vegetation density to such an extent that sedimentation and erosion from marsh surfaces are affected? How does increasing tidal inundation due to sea-level rise alter the allocation of plant photosynthate (chemical product of photosynthesis) to roots, thereby changing the rate of marsh elevation increase? How does deposition of wrack (marine vegetation) carried into the marsh by storms affect the composition of the plant community? In some cases, coastal retreat is accelerated; in other places, little change in the coastal landscape is observed. What combination of rate process changes cause tidal marshes to be converted to mudflats or subtidal zones? How will sea-level rise influence the morphodynamics of the large lowland river deltas around the world and of river-dominated coastal zones? Could such effects lead to rapid wetland drowning and abrupt landward shifts in shoreline? Other consequences of sea-level rise include salinization of aquifers and estuaries, and damage to coastal infrastructure and economies, particularly fisheries and agricultural production.

A major challenge facing Earth surface process science is to couple biological and ecological processes with physical processes to produce predictive models of coastal landscapes that consider the biological effects on flow and sediment transport and to understand how coastal ecosystems are altered as function of morphology, flow, and sediment transport. Without such coupling it will be impossible to approximate the behavior of these systems. The Coastal Zone is an area in which interaction among hydrological, geological, and ecological scientists is especially crucial (see also Section 2.6).

Understanding Limits of Landscape Resilience to Change

The conditions that lead to landform transformation are difficult to predict, and few state changes have been well monitored. Comparison of neighboring landscapes affected by differing climate, tectonic, and volcanic events, or alterations of vegetation by anthropogenic disturbances, provide "natural experiments" that can be exploited through field and modeling studies of the changes. River channels and coastal sedimentary landforms tend to be more sensitive than hillslopes or glaciers to such alterations, and provide some of the most convenient examples for study on the time scale of instrumental and isotopic records.

Characterizing climate in terms that are relevant for process studies at Earth's surface is also needed. Whereas many mechanistic explanations of these processes are event-based (for example, a rainstorm or windstorm of finite duration, intensity, and extent), tests of climate models are based mostly on mean climatology. The ability of these models to represent extreme weather events, which have strong impacts on landforms, is not clear. The outputs of most climate and weather models are provided on spatial scales that are too coarse to resolve rain rate at the scale important to landforms and are not organized in a manner useful to understanding landforms, despite the introduction of concepts such as geomorphically relevant climatic statistics more than a half-century ago (see also Section 2.3). High-resolution regional climate models (spatial resolutions from a few kilometers to a few meters), coupled with the surface hydrology and vegetation dynamics and ultrahigh-resolution global climate models (spatial resolutions of a few tens of kilometers), are being developed. Progress in incorporating this type of resolution could support the development and testing of predictive, mechanistic descriptors of land surface response to climate change.

2.8 HOW WILL EARTH'S SURFACE EVOLVE IN THE "ANTHROPOCENE"?

Humans have changed the face of Earth throughout history and pre-history. Through agriculture and urban development, humans have altered land cover, promoting soil erosion (Box 1.1) and interfering with hydrologic and biologic processes (Figure 2.23). Dams and levees have been built for flood control, water supply, and hydroelectric power, while interrupting sediment movement along rivers and causing loss of habitat and biodiversity. Chemicals have entered water and soil through industrial and agricultural practices. Human activity has also been linked to our warming climate over the past several decades, and this warming, in turn, affects Earth's surface processes. Humans have changed the surface of the Earth more drastically over the past 50 years than at any time in history. These effects may increase with a growing global population that is projected by 2050 to surpass 9 billion and possibly reach 10.5 billion inhabitants (United Nations, 2009). The environmental impacts of human population growth and the accompanying resource consumption and disposal have become so pervasive that the term "Anthropocene" has emerged in the scientific literature to signify a new geologic era (Crutzen and Stoermer, 2000), with the contemporary global environment dominated by human activity (Steffen et al., 2007; Zalasiewicz et al., 2008). Clearly, even identifying "natural" landscapes on the planet is now difficult (see Box 2.10).

What must we do in order to understand, predict, and respond to rapidly changing

FIGURE 2.23 Agricultural (khet) terraces in the Garhwal Himalayas, Nepal. The use of fertilizers for agriculture has dramatically increased the input of biologically available nitrogen into terrestrial ecosystems. SOURCE: Photo courtesy of Richard Marston, Kansas State University.

landscapes that are increasingly altered by humans? This question is among the most pressing challenges of our time, and its scientific component falls squarely within the purview of Earth surface scientists. We now need to build conceptual and numerical models that account explicitly for human-process interactions on Earth's surface, even if such models may have inherent limits relative to our desire for quantification and prediction. Models should strive to define the limits of predictability and the degree to which certain problems are unpredictable. In addition, because theories for Earth's surface systems are developed largely for natural landscapes and because such landscapes are increasingly rare, we also need new theories based on enhanced understanding of human-landscape interactions. Such knowledge is important in developing the tools needed for adaptive management and for guiding decision making, especially in the face of uncertainty and change. Because much of landscape restoration targets a return to "natural" conditions, clearer understanding of the human modification of

Earth's surface would also enable Earth surface scientists to inform resource managers when this goal is possible or sustainable at reasonable expense (see also Section 2.9).

Significant advances have been made over the past decades in defining the extent of human impacts on Earth's surface. Rapid development in tools, particularly lidar and satellite remote sensing as outlined elsewhere in this chapter, has also provided unparalleled opportunity to acquire new data for examining the significance of human alteration of landscapes (see Box 2.4; Appendix C). These data can facilitate development of predictive frameworks for human-landscape systems if knowledge across disciplines also can be integrated with new models and approaches. Recent workshops supported by NSF have facilitated collaboration between geomorphologists, geochemists, and ecologists working on the dynamics of Earth's surface that include human impacts. They have also enabled the establishment of common areas of research in which natural and social scientists are beginning to integrate social and environmental analysis (e.g., Vajjhala et al., 2007; Nagel et al., 2009). In this context, the following paragraphs highlight three central areas of research that are poised for substantial progress over the coming years and identify the challenges to, and high-priority needs for, making advances possible.

Long-Term Legacy of Human Activity

The dominance of humans in shaping some modern Earth surface environments is clear (Figure 1.3). Over decadal and centennial scales, the impacts of human alteration of Earth's surfaces, especially changes in land use, will likely exceed those of climate change that has captured global attention. Yet, the expected magnitude and ramifications of these impacts are not well understood or quantified. For example, how do we separate the effects of human influences from those of other, natural processes? How close are human-impacted landscapes to thresholds for ecosystem collapse or the onset of rapid erosion or mass wasting, especially under global climate change (see also Section 2.7)? How do we predict multiple and/or compound effects of human activities? Which regions are particularly vulnerable to change? What might be the impacts of these changes on human society?

Understanding the legacies of human impacts on landscapes remains an urgent research need especially because the information would provide society with options for adaptation. Meeting this need requires accelerated field and remote-sensing studies to identify the key characteristics of landscapes affected, in addition to using historical records, survey maps, and data from paleoenvironmental studies for documenting and gauging landscape change over long periods. Emphasis is needed in sensitive environments prone to rapid change (see also Section 2.7). These include coastal and urban areas where human populations are concentrated, as well as polar, mountain, and arid regions where increasing human impacts pose pressing issues of resource use. Developing predictive models of system response to anthropogenic changes also requires reliable

BOX 2.10
The Illusion of "Natural" Landscapes

Many landscapes around the world reflect a long history of human presence. This presence is obvious in some cases where rivers, for instance, have been channelized, in constructed environments where hillsides have been intensively terraced (Figure 2.23), or where water-intensive crops have been grown through irrigation in normally arid zones. In other areas, human influences are not as apparent because the initial alterations occurred centuries or even millennia ago (see figures below). For example, catchments of the Pacific Northwest region of the United States contained abundant large wood at the time of European settlement, but so much wood has been removed to reduce flood hazards and improve navigation in rivers that people now tend to perceive wood in rivers as "unnatural" and requiring restoration. In the eastern United States, studies have also shown that before European settlement, small anabranching channels within extensive vegetated wetlands accumulated little sediment but stored substantial organic carbon. However, tens of thousands of water-powered mill dams built in the seventeenth to nineteenth centuries raised the regional base levels of stream channels, inducing sedimentation behind the dams and burying the wetlands. Subsequent breaching of the dams caused channel incision and accelerated bank erosion that resulted in the modern incised meandering streams. The concept that these modern, meandering channel forms represent "natural", or ideal, conditions has guided theory development in fluvial geomorphology for decades, but has only recently been proven incorrect with investigations of the seventeenth to nineteenth century human influence in these areas. Returning landscapes to their original, natural conditions underlies much of the practice of stream restoration (see Box 2.13). Deciphering what is truly "natural" on an Earth that is not static presents challenging scientific, sociocultural, and policy questions for Earth surface scientists (see also Section 2.9).

Grand Challenges in Earth Surface Processes

Top: La Saone River through Lyon, France, exemplifies a channelized urban river. *Middle*: Portland Creek in the Coast Ranges of Oregon is a product of centuries of removal and subsequent reintroduction of wood by management. *Bottom*: The incised meandering streams characteristic of the mid-Atlantic region of the eastern United States were once small anabranching channels that existed within extensive vegetated wetlands; photo shows Western Run at the upstream end of a mill pond. SOURCES: *Top and middle* used with permission from Anne Chin, University of Oregon. *Bottom*: from Walter and Merritts (2008); reprinted with permission from AAAS.

quantitative data about human impacts on processes and process rates. Our capability to collect large datasets through surveys (for example, with airborne laser swath mapping [ALSM]) or field monitoring (using networks of wireless devices) will improve our ability to address the long-term human impacts on landscapes over the various time scales involved; however, significant challenges remain in data analysis and interpretation. A need exists for increased collaboration between Earth surface scientists and geospatial scientists with expertise in the application of existing and emerging geospatial and remote sensing technologies. Lack of understanding of the human responses to anthropogenic landscape change also limits development of the integrated models needed by society. This limitation points to the need to strengthen interactions between the natural and human sciences.

Complex Interactions within Anthropogenic Landscapes

Interactions in human-impacted landscapes have been examined recently within a framework for complex environmental systems (Pfirman and the AC-ERE, 2003). This framework recognizes the interconnectedness among Earth's surface components, the non-linearity of the relationships between them, the historical conditions and inheritance that relate to local settings, and the range of spatial and temporal scales for examining these systems. For example, although the installation and operation of dams provide many economic benefits, these structures also induce physical, biological, and chemical interactions in rivers that may not have been anticipated (Box 2.11). These interactions often intertwine with social processes, as in the case of the Klamath River of California and Oregon, where dikes and levees have increased water supplies since 1905 but adversely impacted threatened and endangered fish (NRC, 2008). Although management actions undertaken during the dry year of 2001 protected the basin's declining fish population, they reduced the water available in the system and triggered further physical and biological reactions, as well as social conflicts related to agriculture.

Although singular impacts and responses are known in many cases, multiple stressors, their impacts, and consequent interactions make human-influenced landscapes difficult to model or predict. How do social processes influence these interactions? How will the mutual interactions adjust to global climate change? Which environmental processes are most vulnerable to change? How will these changes affect the sustainability of water and biotic resources? To improve predictive capabilities for these complex interactions, focused interdisciplinary field studies that link multiple processes are needed as well as experimental studies and theoretical modeling (see also Section 2.6). Incorporating decision making and social processes in some of these interactions also requires strengthening ties between the natural and social sciences.

Coupled Human–Landscape Dynamics

Traditionally, Earth and environmental scientists have treated humans as external drivers of change in examining human impacts on the environment. For river landscapes, for example, studies have emphasized adjustments in fluvial forms and processes following deforestation, mining, impoundment by dams, and urban development. Knowledge of how landscapes have changed after human disturbances has been important in formulating theories of landscape change and assisting management. Yet the traditional approach is insufficient to capture the interrelationships and feedbacks that characterize landscapes affected by increasing and multiple human-caused stressors. Examining coupled natural and human systems enables integrated understanding of how humans both influence and are affected by natural patterns and processes (Pfirman and the AC-ERE, 2003).

The reciprocal relationship between natural and human systems may exhibit positive or negative feedbacks. In Kenya, for example, where conversion of forests into cropland has degraded soils and decreased crop yield, greater food insecurity has hastened the conversion of remaining forests to agriculture, illustrating a positive feedback that accelerates environmental degradation. Negative feedbacks often involve modification of human behavior and policy that potentially slows or halts the impact. For example, flow regulation in Cypress Creek, Texas, eliminated the variability in flows and eventually threatened the culturally and aesthetically valuable Bald Cypress tree, prompting local stakeholders to reevaluate and adjust policies for dam releases. Historically, degradation of Earth's surface by humans has been accompanied by only weak negative feedbacks on human behavior because humans have proven innovative in their ability to endure their own continued alterations to an environment with little regard to its original or "sustainable" function. Weak feedbacks also result where the causes of the impact originate at different scales and faraway places, as in the case of the impacts of global climate change on Arctic residents. Earth surface science plays an important role in public policy by recognizing these diffuse, and often subtle, influences while there is still time to mitigate them. Understanding coupled human-landscape dynamics is among the epochal challenges for predicting the evolution of Earth's surface in the presence of the growing human population and for developing innovative solutions for environmental management.

The study of coupled natural and human systems has thus far emphasized ecological phenomena, producing an emerging science for social-ecological systems that explicitly incorporates human decisions, cultural institutions, and economic and political systems. Enormous gaps in knowledge remain, however, especially with respect to understanding the impact-feedback loops within geomorphic systems. Social processes influence both the original human impact and the potential responses and feedbacks within landscape systems—in other words, human systems influence whether crops are grown on particular soils or how a city is developed, and human systems need to manage and adapt to the

> **BOX 2.11**
> **Constructing and Removing Dams**
>
> Of all the human alterations of the Earth's surface, the construction of dams and the acceleration of erosion through agricultural and other land clearance activity are among those having the most profound effects on river sedimentation and morphology. Nearly two-thirds of the world's rivers are regulated by dams, levees, and other structures. In the United States alone, approximately 80,000 dams are registered in the National Inventory of Dams, many built in the 1960s. Millions more small impoundments dot the landscape. These structures can collectively store a volume of water equaling almost one year's runoff. The large ones alter the timing and magnitude of river flows and dissolved and particulate material, as well as the locations and pathways of chemical reactions. The dams also compartmentalize our nation's rivers, fragment landscapes and aquatic habitats, and create new reservoirs for sediment and carbon. Globally, dams have reduced the flux of sediment reaching the world's coasts by 1.4 ± 0.3 billion metric tons annually, despite increased quantities of sediment introduced through soil erosion and transported by rivers (2.3 ± 0.6 billion metric tons per year) (see figure below). The diminished transfer of sediment from land by rivers is increasingly important with respect to global denudation and geochemical cycles, the functioning and loss of coastal ecosystems, and the stability and evolution of deltas and depositional environments (see also Box 2.2). Climate change is expected to accelerate and confound these impacts, as melting glaciers (see also Section 2.7) increase the potential need for water storage by dams, and sea-level rise further threatens the viability of coastal living space. Quantifying these potential impacts is hampered by the limited availability of data for sediment loads. Since many reservoirs are now approaching their limits of usefulness because of sedimentation, and as environmental impacts and the associated loss of ecosystem services are increasingly recognized, decommissioning and removing dams and other structures also bring new opportunities for ecosystem restoration (see Section 2.9).

environmental changes, such as accelerated erosion, that may be caused by these activities. Thus, a primary need in understanding the coupled system is the exchange of analytical perspectives between Earth surface scientists and social scientists, including cognitive-behavioral scientists, economists, political scientists, sociologists, and human geographers. Anthropogenic Earth surface changes also need to be quantified in terms of how they, in turn, affect humans directly or indirectly. For example, accelerated soil erosion (Box 1.1) can be quantified in terms of how it affects the water-holding capacity of the soil profile, and thus plant and crop production, and predicted over time scales that can be expected to be valued by humans. These changes can also be translated into the language of ecosystem services, human perceptions and valuation, and the extent to which predictable thresholds exist that would trigger feedback responses to accelerate or decelerate the original impact. Because policy and institutional processes are the key instruments that society has for effect-

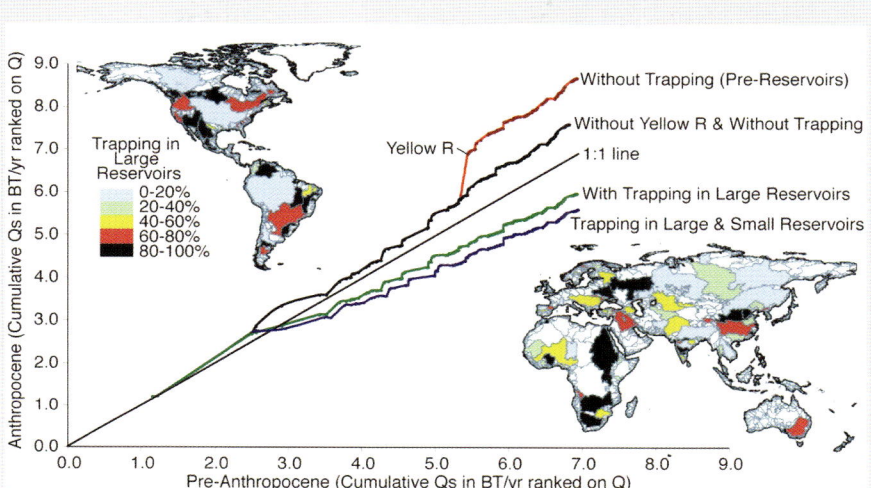

Comparison between pre-"Anthropocene" and modern ("Anthropocene") sediment loads using data from 217 rivers before and after dams were constructed. Data are presented as cumulative curves ranked by decreasing river discharge. Two curves above the 1:1 line (representing no human influence) indicate the increased sediment yield caused by deforestation (soil erosion) and other human activity. Two curves below account for the impact of sediment sequestering in reservoirs. Inserts show the basinwide trapping of sediment by large reservoirs, reported in Vorosmarty et al. (2003). SOURCE: Syvitstki et al. (2005); reprinted with permission from AAAS.

ing necessary feedbacks, scientific frameworks for coupled human-landscape systems would also require linkage to policy mechanisms. Successful transmission of scientific information is required to guide policies for mitigation and adaptation, pointing to the need for research findings to be articulated in forms that are useful for policy making, as well as the identification of pathways for transmission of this information to policy makers.

Science is far from developing a general theory of coupled human-natural systems, even though such a theory may offer the potential to slow or reverse environmental degradation and improve human well-being. Because such a theory would include knowledge of societal perceptions of environmental impacts and the ability and willingness of societies to react to these changes, much focused inductive and empirical work is required to investigate these interacting processes in a range of Earth surface environments, and in societies at various levels of institutional development and types of organization.

2.9 HOW CAN EARTH SURFACE SCIENCE CONTRIBUTE TOWARD A SUSTAINABLE EARTH SURFACE?

As human activity became widespread and intensive some 5,000 years ago, the surface of Earth was—as it continues to be—altered in many ways (Section 2.8, Figure 1.3). With increasing scientific understanding of the causes and cumulative, long-term effects of human-induced changes, a consensus has emerged that at least some of these disrupted or degraded landscapes can and should be "restored" or "redesigned." The meaning of the term "restored" is not agreed on with precision, but it generally refers to a set of strategies to reestablish processes that functioned before the most recent and intense anthropogenic disruption. Where this is not possible, a more limited set of natural processes may be reestablished, or a fuller set may be reestablished at a smaller scale than the original case. Restoration is often incorporated into engineering strategies for making landscapes safer for human habitation (such as coastal barrier reinforcement or flood risk reduction).

Landscape restoration may involve river channels and floodplains (river restoration), entire ensembles of hillslopes and small channels (watershed restoration), or coastal wetlands and waterways. Landscape restoration is typically managed by coalitions of engineers, biologists, and planners, but as restoration activities become more sophisticated they will include a broader range of Earth surface scientists. They will also involve social scientists, planners, and practitioners, as in the development of integrated assessment tools to guide decision making. The design of landforms (in the case of rivers, floodplains, and tidal marshes) or land conditions (in the case of watershed restoration) that will function within a predictable, preferred range has had some success. The real challenges emerge in designing landforms that interact with the desired biological processes and communities required to establish sustainable ecosystem complexity (Box 2.12). In fact, biotic processes are central to the evolution of many landforms (see also Section 2.6). Hence, the design and monitoring of restored landscapes and their ecosystems should both rely on and add to emerging models of interacting biological, physical, and chemical processes in landscape evolution. Furthermore, the long-term cost of a project will be minimized and the likelihood of its sustainability will be maximized if the original design lies within the stable range of landscape fluctuation and if it respects long-term trends in landscape evolution.

To this activity, research in Earth surface processes brings knowledge of how major environmental drivers (such as water regime or sediment supply) scale the size and shape of landforms that are treated as the "reference conditions" for restored features. Thus, specific research challenges arise about controls on the size and geometry of redesigned rivers, tidal creeks and inlets, deltas, and beaches and how they are controlled by the expenditure of flow energy and by sediment supplies. In other cases, where landforms originally existed in sediment-starved environments or were dominated by intense biogenic drivers such as large wood or riparian vegetation, little theory exists at the present time for how such landscapes might be recreated or what the alternative sustainable states are. In these cases, the need is for extensive field investigations to support new theories of the nature and functioning of biogenically reinforced landscapes that would retain sediment and chemicals and release them slowly into water bodies.

Research in Earth surface processes also focuses on the transport processes that redistribute sediment and chemicals, and interact with biotic processes such as seed dispersal, plant survival, and colonization of aquatic substrates by woody plants, algae, and macrophytes. Attempts to modify the sediment and carbon storage in marshlands and soils require new field studies of the accumulation of organic sediments. Where human effects involve the disruption of sediment supplies, restoration entails reducing excessive sediment supplies from disturbed areas using principles of hydrology, geomorphology, plant community regeneration, soil profile enhancement, and engineering. In other cases, perturbations such as the removal of dams or the augmentation of beaches involve increases in sediment supplies to landforms, and the effects of these perturbations need to be predicted through advances in the study of mixed-grain-size sediment transport that may transform or even engulf the landform itself. The nation's gradually expanding program of dam removal from rivers (now ~50 projects per year; see also Box 2.11) adds urgency to the need for research into sediment management along rivers. In most dam removals, the major problem to be solved is not the removal of the dam or the restoration of flows, but the management of impounded sediment. Strategies for stabilization, removal and disposal, or release of this sediment involve large costs and ecological and societal consequences, and pose research challenges for Earth surface scientists and engineers involved with sediment transport and channel mechanics (Box 1.1).

Earth surface processes research also illuminates the long-term context of restoration plans. Restoration of landscapes requires designs that will survive long-term conditions such as highly stochastic sediment supplies, regional land-use changes, the relaxation of some landscapes from deglaciation, sediment starvation, crustal deformation, and sea-level change. Studies of these phenomena clarify the limitations on restoration and, especially, the role of time in allowing ecosystem complexity to develop. In other situations, the disruption of flows and sediment supplies or the deepening of coastal waters through land surface subsidence renders restoration to a preferred state with desired biological characteristics impossible on time scales of human interest. Research into the Late Cenozoic evolution (the last several million years) of sedimentary landforms and their ecology and biogeochemistry provides a basis for understanding this long-term context and sustainable environmental range of landscape restoration (see also Section 2.1).

At the same time, restoration projects provide field-scale experiments within which the evolution of landforms or soil profiles can be studied with a resolution and degree of simplicity that are rarely possible in nature. Initial conditions are well known and usually simpler than natural landscapes, and processes occur at a natural scale as opposed to miniaturized laboratory conditions. Often, variables such as sediment supplies to beaches,

> **BOX 2.12**
> **Coastal Wetland Restoration**
>
> Many tidal marshes have been converted to non-wetland habitat by dredge-and-fill activities, nutrient loading, groundwater extraction, contaminated runoff, fire management, and invasive species. Coastal wetlands have declined worldwide by 0.5-1.5 percent per year over the past few decades. In Louisiana, losses in coastal wetlands have risen to ~60 km^2 per year or nearly 18 hectares (~45 acres) per day (Barras et al., 2003). Sea-level rise and crustal subsidence also lead to loss of wetlands, but the destruction of wetlands by direct and indirect human activities is of much greater consequence (see also Box 2.2).
>
> The rate at which coastal wetlands continue to be filled and drained belies their ecological and economic importance as wetlands. Wetlands are among the world's most productive ecosystems. They provide habitat, food, nurseries, and refuge for fish and shellfish. Wetlands store and cleanse water, prevent coastal erosion, and reduce storm surges. In the 1960s, public awareness of the importance of wetlands led to policy changes intended to slow wetland loss by encouraging wetland mitigation and restoration (e.g., Section 404 of the Clean Water Act in 1972).
>
> Coastal wetlands develop at the interface between land and water. Their very existence results from a tenuous, but enduring mass balance of mineral and organic sediment that is imported, exported, and generated in situ in locations that lie approximately at mean sea level and are simultaneously subject to the eustatic rising of sea level, the warping of Earth's crust as a result of the last deglaciation, tectonics, and sediment loading (often altered by human activity), and anthropogenic effects such as groundwater extraction. Many coastal wetlands have very low topographic gradients (see figure below), so that differences of only several centimeters in salt marsh elevation strongly influence vegetation patterns and ecological and biogeochemical processes.
>
> Restoration of coastal wetland ecosystems requires the land surface to be sufficiently high in the tidal range that vegetation is not flooded constantly, but also low enough that the energy and sediment supply provided by regular flooding is achieved. Vegetation exerts strong controls on water movement and sediment deposition and erosion that shape the marshes and tidal creek networks. Until recently, only the effects of physical processes on biotic processes were considered in restoration of coastal marshes. This approach focused on contouring sites, planting native species, redirecting sediment dispersal at river mouths, and reintroducing tidal flushing. The outcomes of such efforts, however, were natural vegetation patterns, but not functionally diverse marshes (e.g., Tijuana River Estuary, California; Barn Island Wildlife Management Area, Connecticut).
>
> Reestablishing functional, as well as structural, characteristics of coastal wetlands will require fundamental advances in understanding the coupling of ecological and morphodynamic processes at a range of scales (Fagherazzi et al., 2004). Within individual marshes, emphasis needs to be placed on incorporating

base levels of wetlands, and flows to rivers are manipulated in ways that allow quantitative expectations to be developed (Box 2.13). Observations in the field of the effects of real trees, root zones, material properties, and sediment supplies, for example, cannot easily be reproduced indoors. In other cases, animals are excluded from large areas or small plots, or climate is controlled through watering or soil profile warming and fertilization. Field experiments topographic and hydrologic spatial heterogeneity and temporal variability for distributing water and sediments in ways that support compositionally and functionally diverse wetlands. At the watershed and landscape scales, there is a need for greater knowledge about wetland diversity within watersheds, the proportion of uplands to wetlands, and the physical and ecological processes connecting wetland and upland ecosystems (NRC, 2001b).

Increasing rates of sea-level rise and projected increases in the frequency and intensity of tropical and extratropical storms will lead to greater threats to coastal shorelines, wetlands, and developed areas. Management and restoration of coastal landscapes will be dramatically complicated by these effects because more than half of the U.S. population lives on coastal landscapes. Accurate predictions of how coastal landscapes will respond to sea-level rise and disturbance due to storms will provide coastal managers with information that allows them to reduce the risk of potential hazards rather than responding to destruction that has already occurred (e.g., Hurricane Katrina).

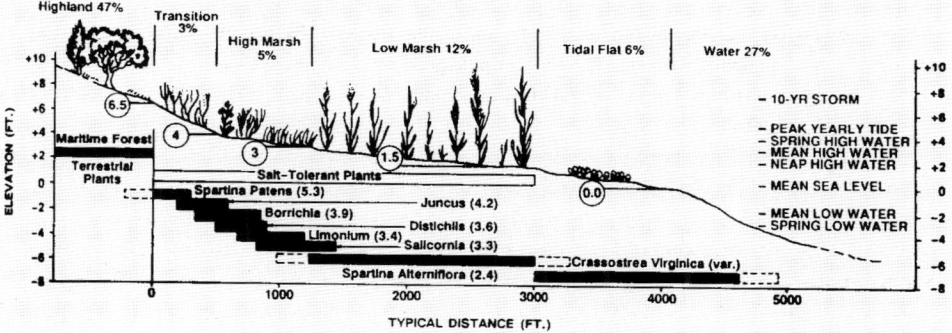

Composite wetlands transect for Charleston illustrating the approximate percentage occurrence and modal elevation for key indicator species or habitats based on results of 12 surveyed transects. Minor species have been omitted. Elevations are with respect to 1929 NGVD (National Geodetic Vertical Datum), which is about 15 centimeters lower than current sea level. Current tidal ranges are shown at right. SOURCE: Titus (1988).

with replication are often possible within the confines of restoration projects. Research opportunities include the high-resolution testing of predictions of the stable scale or range of some landforms, as well the trajectory of other features that may not equilibrate. Recently developed geochronological and geophysical tools and geochemical tracers enhance the nature and level of detail of studies that are now possible of both the mechanisms and the

> **BOX 2.13**
> **River Restoration**
>
> What shape and gradient should a river have in order to behave in a more or less predictable manner consistent with human designs? How much water should flow through it? What plants and animals should live in its waters and on its banks? In the United States, at least $1 billion is spent on river restoration annually, and the number of restoration projects is growing exponentially (Bernhardt et al., 2005). The environmental and economic motives for restoration vary with geographic region and history of environmental impact. Restoration practice has outpaced scientific investigation, in that many of the basic scientific questions relevant to stream restoration have yet to be answered or even investigated. Better understanding of the processes that determine river evolution and response to environmental change, for example, could improve our ability to design sustainable river systems that provide ecosystem services that people desire.
>
> The term "river restoration" begs the question: Restore to what state? "Restoring" a river usually means returning it to a close approximation of its condition prior to a disturbance. Yet understanding a river's past and present states is often complicated, requiring the use of a broad set of tools and disciplinary approaches that include many Earth surface processes. Rivers are dynamic features that adjust to changes in climate, hydrology, vegetation, and human activities. Increasing sediment input to a river—by logging of neighboring forests, for example—may increase erosion of stream banks. Left unchecked, stream bank erosion can cause the river to become wider and shallower and—in severe cases—to migrate erratically and flood more frequently. Wastewater discharge or runoff from agricultural fields can change the chemical load of rivers as well as their habitability for different species. River restoration projects not only take stock of past and current processes that control the long-term stability of streams, but also attempt to predict what modifications will best achieve the project's goals.
>
> In addition to the scientific framework, rivers are restored within particular constraints of regulatory goals, economic needs, and public preferences that vary geographically. Projects are guided by a variety of federal, state, and local regulations that oversee land use, water quality, and flood control, for example. Regulations also often interact with economic incentives that drive decision making. In one case, after determining that fish passage upgrades and maintenance costs would exceed the value of Oregon's power-generating Marmot Dam (see figure below), the dam was decommissioned and removed in 2007. The dam breaching released 100,000 m^3 of accumulated gravel and sand into the river downstream within the first 48 hours (Major et al., 2008). Although the full consequence of this added sediment load downstream over the coming decades cannot yet be predicted quantitatively, the decade of research that preceded dam removal made accurate predictions of some of the impacts of sediment release and aided the decision-making process for decommissioning and removing the dam.

rates of change to be expected from managed or reconstructed landscapes (Boxes 1.2, 2.1, 2.4, 2.6; Appendix C). Thus, landscape restoration provides an opportunity both for learning and for service by the Earth surface processes research community.

In addition to restoring and redesigning landscapes, society needs Earth surface scientists to guide decisions regarding the functioning and evolution of the surface, while

Aging infrastructure, combined with shifting technological practices and environmental values, has contributed to environmental restoration efforts that include dam removal and stream restoration. Here, the 15-meter-high Marmot Dam on the Sandy River is destroyed in July 2007. This hydroelectric dam was built in the early twentieth century, but was no longer in operation. Sediment stored in the reservoir to the level of the dam breast was rapidly mobilized once the dam was breached, leading to a sudden pulse of sediment downstream. Nevertheless, salmon migrated upstream past the dam site within several days of its breaching. SOURCE: Portland General Electric.

recognizing uncertainty and gaps in scientific knowledge. In human-dominated regions, for example, landscape-modifying processes are often simplified by engineering and other forms of resource management, but re-naturalizing the processes is not simple. Complex historical cascades of decisions have often been made about rights to resources of land, water, and minerals and multiple layers of interests have solidified in the status quo. If this history has resulted in the degradation of resources or ecosystem services, or even the viability of human living space, the complexity of human interests creates resistance to unraveling the damage. This difficulty is compounded by the complexity of the physical, chemical, and biological processes themselves, which makes it difficult to predict the results of changing resource utilization or other societal action. Yet, society needs Earth surface scientists to construct models for assessing the probable effects of various possible restoration actions. The models may initially be crude and may require sequential elaboration and refinement. Learning how to develop such assessment tools in a politically sensitive, useful way requires that natural scientists studying Earth's surface processes collaborate with social scientists who study all of the other processes that affect the surface, as well as with practitioners in industry, engineers, and planners.

Landscape restoration is a complex, high-priority goal for many researchers, practitioners, policy makers, and the public, but only recently have these different communities begun to examine together the legacy of past restoration efforts in the context of new data from specific regions of the Earth's surface. Earth surface scientists have vital contributions to offer as restoration activities are carried toward quantitative models and predictions. Just as importantly, Earth surface scientists are in position to guide future decisions through inclusion of understanding of human-process interactions, even while recognizing uncertainty and limits of predictability. Such guidance may enhance the goods and services that Earth's surface provides to society and thereby produce enormous economic benefits, as well as aid in developing a sustainable living surface for the next generation.

CHAPTER THREE

Four High-Priority Research Initiatives in Earth Surface Processes

Although each of the nine grand challenges in Earth surface processes embodies significant prospects for research advances, the committee suggests that these challenges can be collected into four major research initiatives (Figure 3.1). The idea of "research initiatives" is further developed in Chapter 4, but the essential proposal is that these initiatives are timely, high-priority research areas that are rich in scientific merit and will potentially transform the field of Earth surface processes. They will require new interdisciplinary approaches, including the integration of theories, models, and tools. The four high-priority research initiatives are (1) interacting landscapes and climate; (2) quantitative reconstruction of landscape dynamics across time scales; (3) the coevolution of ecosystems and landscapes; and (4) the future of landscapes in the "Anthropocene". These initiatives are poised for launch by researchers but will require coordinated efforts to develop the new intellectual collaborations needed among communities of scientists. Working with new combinations of scientific approaches, tools, and models will achieve transformative advances in the study of Earth surface processes. The scientific objectives, challenges, and intellectual collaborations needed for these initiatives are outlined below. Chapter 4 suggests mechanisms to support development of the initiatives.

3.1 INTERACTING LANDSCAPES AND CLIMATE

Climate is thought to play a critical role in driving the flux of solutes and mass across landscapes and through ecosystems during weathering, mass transport, erosion, and deposition. Topography, vegetation, biogeochemical cycling, and snow and ice cover also influence climate locally as well as globally and over long time scales. Yet much remains unknown about the links and interactions between Earth surface processes and climate. The principal goal of a major research initiative in the area of climate-landscape interactions is to develop a quantitative understanding of climatic controls on Earth surface processes and landscape influence on climate over time scales that range from individual storm events

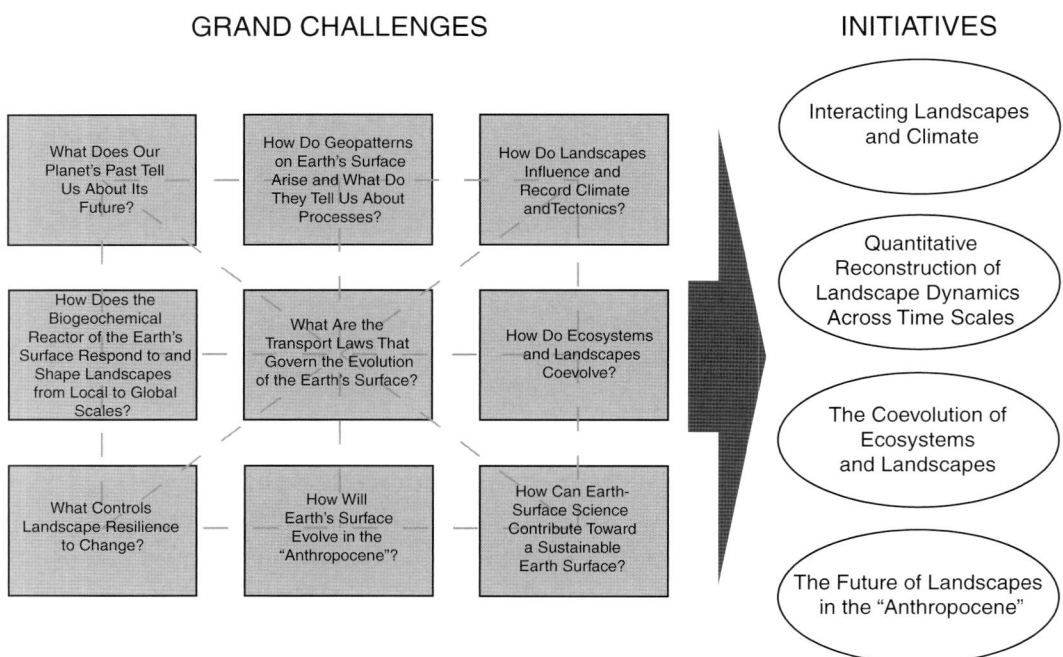

FIGURE 3.1 Conceptual diagram illustrates the relationship among the nine grand challenges and four high-priority research initiatives. The nine grand challenges are interconnected at many intellectual and technical levels (schematically represented by the dashed lines; see Chapter 2 for details). Rising from the fusion of the nine grand challenges, the four high-priority research initiatives are particularly apt to advance understanding and promote interdisciplinary collaboration in Earth surface processes.

to millennia. Longer-term interactions among climate, topography, and tectonics are the subject of a complementary initiative (see Section 3.2). Success with this research initiative has the potential to transform our understanding of the powerful role of climate in shaping and transforming the Earth's surface and the feedbacks between surface change and climate. Such understanding is essential to efforts to predict landscape response to land use and climate change and to evaluate quantitatively the efficacy of various management plans and mitigation strategies.

The primary science objectives for this initiative include the following:

- Development of theory for the interactions among topography, land cover, and global, regional, and local climate that determine the biogeochemically and geomorphically significant attributes of climate (for example, ground-air thermal histories, precipitation, runoff, winds, and waves).

- Development of geomorphic transport laws that explicitly account for climate (probability distributions of temperature, precipitation, runoff, winds, and waves) and incorporate interactions with biota, including theories for river and glacier incision; production, transport, routing, and deposition of sediment; and geochemical processes.
- Monitoring, experimentation, and modeling of climatic controls on the weathering of rock and soil and their influence on physical erosion rates and vice versa. Climatic effects include both the direct effects of characteristics such as air or ground temperature and moisture flux and the indirect effects such as primary production and microbial processes, and incorporate the influence of land surface attributes and mass fluxes on biogeochemical cycles.
- Study of the feedbacks between global and regional climate and (1) the operation of terrestrial carbon reservoirs (including the role of land use and climate change, fire, erosion, and deposition), and (2) the controls on atmospheric dust concentration (including the evolution of soil composition, soil moisture, topography, and land cover in source regions).
- Modeling and monitoring of landscape evolution under diverse and varying climatic conditions; identification of climatic signatures in landscapes; evaluation of thresholds of landscape response and the limits of resilience.
- Development of theories for subglacial hydrology, basal sliding of glaciers, subglacial sediment deformation, and ocean-ice interactions at ice-sheet margins—processes critical for evaluating rates of glacier melting and potential thresholds of ice-sheet collapse.
- Improvement of the coupling between surface processes and existing climate models, explicitly incorporating the effects, feedbacks, and conditions outlined above.

Collaboration between Earth surface scientists and atmospheric scientists is the primary need for developing and advancing this initiative. Recent National Science Foundation (NSF)-supported workshops (Galewsky and Roe, 2008) have brought climate scientists together with geomorphologists, for example, to begin to identify key questions and collaborative research opportunities for Earth surface processes. Because many new research questions in this area clearly cross boundaries between atmospheric science and Earth surface science, improved communication between these communities is essential for significant progress. Increasing the visibility and understanding of Earth surface science within the atmospheric sciences community, and vice versa, through analysis of common scientific objectives is important to help establish these links.

3.2 QUANTITATIVE RECONSTRUCTION OF LANDSCAPE DYNAMICS ACROSS TIME SCALES

One of the paradoxes of Earth's surface is that locally it can evolve and change abruptly, with dramatic consequences for humans and ecosystems, yet overall it may evolve slowly enough that observations on scales of human lifetimes are insufficient to characterize more than a small slice of its dynamics. By greatly expanding our time horizon, we may gauge important but rare events and perceive dynamics that emerge only on time scales of thousands to tens of millions of years or more—for example, the interplay of surface processes and tectonics that links the landscape to the Earth's interior. This major research initiative is focused on developing a bridge in time from instants to eons, using quantitative reconstruction of landscape dynamics across time scales based on emerging high-resolution information recorded in extant landscapes and the sedimentary record.

Discoveries highlighted in this report show that the world we see today reflects evolution over a wide range of time scales, and that the longer the time scale, the stronger is the coupling between the Earth's surface and its interior. Together with new techniques for dating and measuring features and processes at the Earth's surface, new conceptual models of how tectonic and climatic effects come into play at different time scales open the way to building a unified view of the landscape across the full range of time scales on which it evolves. A confluence of interest has developed with regard to how the surface evolves on increasing time scales and how to mine the long-term archive of preserved events for information on the variability of surface dynamics, and in particular, the frequency of rare but potentially catastrophic events.

Thus, the goals of this initiative are (1) to understand how the interplay of tectonics, climate, biota, lithology, and surface processes creates landscapes from human to planetary time scales, and (2) to extract information on event frequency and rates of evolution from landscape archives over this range of time scales. Reconstructed time-sequence evolution of the Earth's surface will be used (1) to test and inspire models that couple tectonics, climate, biota, and landscape evolution; (2) to constrain the frequencies and causes of rare but important surface events; and (3) to provide baseline information on pre-human landscapes as a guideline for restoration and management.

The primary science objectives for this initiative include the following:

- Improving methods for quantitative reconstruction of past Earth surface states (terrestrial paleoclimate, hydrology, soil characteristics, sediment transport and erosion rates, subsidence or uplift rate) from the fragmentary record of landforms, paleobotany, geochemistry, paleo-soils, and sedimentary deposits.
- Developing detailed paleoclimate, tectonic, and sedimentary records of abrupt changes in Earth surface processes and of landscape resilience over long time

scales to understand the tolerable limits of stochastic variability within different geomorphic systems.
- Developing and testing quantitative predictive models for the Earth surface system with explicit coupling across time scales. Important focus areas for improvement of current models include the use of realistic crust and mantle rheologies; investigation of possible coupling to mantle convection; modeling glacial erosion and transport; and coupling to biogeochemical cycles.
- Developing and improving technical capabilities and data collection in near-surface geophysical methods (for example, high-frequency seismic reflection, ground-penetrating radar, and electrical resistivity profiling) to image and measure Earth's near-surface structure and physical properties in three dimensions.

The essential intellectual collaborations for this initiative to succeed are those between researchers with knowledge of deep- and surface-Earth processes. A particular challenge may be to coordinate collaboration among industry, engineering, and academic researchers that is essential to facilitate the exchange of knowledge, tools, and models among these communities. The identification of critical, complementary research objectives may initiate some of the needed interactions.

3.3 THE COEVOLUTION OF ECOSYSTEMS AND LANDSCAPES

Rapidly growing interest in research at the interface of biota and Earth surface processes—the life-landscape connection—is giving rise to the emerging fields of geobiology, biogeomorphology, and ecohydrology, among others. With new ways of measuring how the living and nonliving surfaces co-organize and an increasing ability to link biotic processes and landscape evolution, the opportunity exists to forge a new understanding of the coevolution of ecosystems and landscapes and to address pressing problems of future environmental change.

This initiative could lead to the transfer of basic concepts and theories on physical systems to ecological research, and to the use of ecological principles to guide coupled ecosystem and landscape modeling. A grand goal is to build the capability to predict future coupled ecosystem and landscape states under varying climate and land-use conditions. Developing this predictive capability is of importance for a number of societally relevant issues, such as the future states of rivers and their ecosystems in response to anthropogenic activity, the evolution of tidal marshlands under sea-level rise, or the erosional stability of landscapes and terrestrial carbon reservoirs in response to increased seasonal thawing. As in global climate modeling, the records stored in paleoecologic, geomorphic, and paleosol archives can serve as tests for modeling. Such shared model building will identify, in turn, key field datasets and processes that are in need of focused studies.

The primary science objectives for this initiative include the following:

- Improvement of theory and observations that relate spatial patterns and dynamics of biota to landscape setting (topography, hydrology, and geology) for given climatic conditions.
- Development of models that incorporate both the geomorphic transport laws and the requisite biogeochemical equations to account mechanistically for the role of biota.
- Development of landscape evolution theory that includes the effects of biota (and its possible coevolution with landscape).
- Development of models to predict the coevolution of climate, biota, and landscape processes under a scenario of increased greenhouse gases.
- Development of observations and models for the interaction of biota with stream channel and floodplain morphology and dynamics.

Significant advances in this initiative depend on development of sustained, long-term collaborations between the biological and Earth surface sciences. Recent workshops that have helped to establish some of the needed interactions (e.g., the Binghamton Geomorphology Symposium on Geomorphology and Ecosystems; see also Chapter 2). Integrating ecology, geology, hydrology, and atmospheric science to examine biological-physical feedbacks that shape landscapes could be linked to ongoing efforts at the 26 Long Term Ecological Research (LTER) Network sites in order to foster interdisciplinary approaches (see also Chapter 4). An important opportunity for potential coordination of these types of interdisciplinary efforts was announced early in 2009 by the NSF through Dear Colleague Letters: Emerging Topics in Biogeochemical Cycles[1] and Multi-scale Modeling,[2] with objectives that include understanding how Earth's biological systems respond to and influence its physical and chemical conditions (see also Chapter 4).

3.4 THE FUTURE OF LANDSCAPES IN THE "ANTHROPOCENE"

Human activity is changing Earth's surface, both markedly and rapidly; these changes are likely to increase with an expanding human footprint across the globe. Over the past several decades, substantial advances have been made in understanding the range and extent of human impacts on Earth surface systems. These advances, coupled with technological breakthroughs, present significant opportunities to develop answers to a fundamental, compelling, and urgent question: How can we understand, predict, and respond to rapidly

---

[1] http://www.nsf.gov/pubs/2009/nsf09030/nsf09030.jsp.
[2] http://www.nsf.gov/pubs/2009/nsf09032/nsf09032.jsp.

changing landscapes that are increasingly altered by humans? Although individual disciplines have tackled different aspects of this question, a focused effort is now needed across disciplines to anticipate and mitigate future impacts, as well as to factor the changes that have already occurred into human decision processes. The overarching goal of this major initiative is to transform our understanding of human-landscape systems—integrated systems characteristic of the "Anthropocene"—and our ability to predict how they might evolve in the future.

This initiative transcends many disciplines spanning the natural and social sciences and engineering. A range of scientific expertise will be necessary to integrate theories, models, and approaches toward predictive capacity for human-landscape systems. The new interdisciplinary knowledge produced will be critical for guiding landscape management and restoration and informing environmental policy. Such guidance will be vital as human population grows and natural resource challenges are increasingly confronted, and as we strive toward a sustainable Earth surface for the next generation.

The primary science objectives for this initiative include the following:

- Improved understanding of the long-term legacies of human impacts on landscapes and quantification of current rates of impacts (e.g., from mining, grazing, deforestation, creation of impervious surfaces, agricultural erosion and pollution, flow and sediment impoundment)—especially in environments that are sensitive to global climate change.
- Development of mechanistic models linking multiple and cumulative effects of human activity.
- Development of integrated models of the complex interactions within human-dominated landscapes, incorporating decision making and human behavior.
- Greater understanding and predictive capacity for coupled human-landscape dynamics.
- Capacity building to anticipate and guide options for mitigating, reversing, and adapting to human-caused landscape change.
- Coordinated collection and database management of sociological and geographical information on land use for incorporation into quantitative models.

Developing this initiative requires new, sustained collaborations among Earth surface scientists and a range of social and behavioral scientists including economists, political scientists, psychologists, sociologists, and human geographers. This range of expertise is necessary to develop the capacity to predict coupled human-landscape dynamics, to incorporate decision-making and human perception and valuation in quantitative models, and to build integrated assessment tools needed by society. Partnership among Earth surface scientists, engineers, practitioners, and planners is also needed to use manipulated landscapes for experimentation and to incorporate feedback mechanisms in scientific frameworks for human-landscape systems. Collaboration with geospatial scientists is additionally necessary to integrate existing and emerging geospatial technologies into modeling efforts, to facilitate data mining and management, and to link remote-sensing studies with field and modeling approaches. Increased collaboration among Earth surface scientists (e.g., ecologists, geomorphologists, and geochemists) and climate scientists is also necessary to address the interactions with global climate change.

Developing intellectual collaborations with social scientists represents a major challenge and a significant opportunity for Earth surface scientists. This area is one in which comparatively little work has been done traditionally, particularly in investigating geomorphic systems. Yet, transformative advances require bridging the real and perceived gaps between the approaches of the natural scientists and those of social scientists. Initial efforts have incorporated social sciences into research at NSF's LTER sites (see also Section 2.6 and Box 4.1), at other environmental observatories (e.g., Vajjhala et al., 2007), and within ecohydrology programs (e.g., in UNESCO). Recent work has also addressed the integration of social sciences into climate-change research (e.g., NRC, 2007b; Nagel et al., 2009). Critical breakthroughs will require accelerating these efforts and initiating new ones that focus on human-landscape systems in which geomorphic processes are central. NSF announced its goal to increase collaboration between the geosciences and social and behavioral sciences in February 2009 through the Dear Colleague Letter: Environment, Society, and Economy.[3] The augmented funding available provides an opportunity for potentially developing this initiative on the future of human landscapes (see also Chapter 4).

3.5 SUMMARY

Interest in Earth surface processes has grown substantially in recent years in response to a confluence of exciting scientific discoveries, greater awareness of the impacts of humans and climate on Earth's surface, new tools and instruments for acquiring data, and increased societal demand for scientific guidance to understand, manage, and restore landscapes. These advances, along with the development of disciplinary programs at NSF that address research at the Earth's surface, have been positive for the field and for society (see also Chapter 4). Nevertheless, the field of Earth surface processes has reached a point at which the imminent research questions cross multiple disciplinary boundaries and require new intellectual collaborations and approaches. The four initiatives outlined in this chapter offer exceptionally promising directions with the potential to lead the field toward major phases of scientific discovery. They are designed to yield fundamental knowledge of past and present Earth surface systems in order to develop predictive capabilities for the future.

---

[3] http://www.nsf.gov/pubs/2010/nsf1003/nsf1003.jsp.

Because of unprecedented rates of change occurring on Earth's surface from the combined effects of human impacts and global climate change, this knowledge is pressing and urgent for both science and society. The four initiatives are suggested to focus the efforts of the research community, as well as those of NSF and other agencies (see also Chapter 4 and Appendix C), toward achieving these goals.

Development of the four initiatives could lead to a new generation of researchers with an increased ability to work in interdisciplinary settings. Rapidly growing numbers of new researchers (faculty and students) within the field underscore a critical need for investigator-driven research opportunities (see also Chapter 4 and Appendix C). Significant progress on the complex problems highlighted in the initiatives will require increasing collaboration among a range of scientists who may not have worked together previously, presenting key challenges to the development and success of the initiatives. The intellectual collaborations needed for each of the initiatives vary; they include interactions between Earth surface scientists and atmospheric and biological scientists, between researchers with knowledge of deep- and surface-Earth processes, and between Earth surface scientists and social and geospatial scientists and engineers. The development and refinement of tools for data collection will require a new level of collaboration between academia and industry. Specific needs in technology, data collection, and the establishment of monitoring networks and community facilities are outlined in Chapter 4, as are specific organizational mechanisms to support development of the four high-priority research initiatives.

CHAPTER FOUR

Mechanisms for Developing Initiatives and Sustaining Growth in Earth Surface Processes

This report has identified nine grand research challenges and four high-priority research initiatives in a rapidly evolving field—Earth surface processes—that is poised to take a vital, international position in predictive, quantitative Earth and environmental sciences. To realize the full promise of the field through development of the four research initiatives, the nation's research structure for Earth surface processes faces challenges to build intellectual and technological infrastructure in three main areas: (1) data collection and distribution, (2) instrument technology, and (3) interdisciplinary collaboration and community building. This chapter suggests mechanisms that could support development of the initiatives as well as sustained growth in the field of Earth surface processes. Some of the mechanisms could be enacted through existing means and as a complement to current efforts by the National Science Foundation (NSF) and other federal agencies to support this field. The first section of the chapter provides a brief review of the mechanisms already in existence at NSF that can support research in Earth surface processes. The next section elaborates on actions and targeted mechanisms for data collection and distribution, technology development, and community building that can accelerate growth in the field. The last section outlines specific ways through which the four research initiatives can be developed, emphasizing the intellectual collaborations outlined in Chapter 3 and the potential means to overcome the challenges of establishing those collaborations.

4.1 FEDERAL RESEARCH FRAMEWORK

NSF, particularly through the Directorate for Geosciences (GEO) and its three Divisions of Atmospheric, Earth, and Ocean Sciences, is a critical source of funding for basic research in all of Earth science. Through the Division of Earth Sciences (EAR), NSF has supported research on a range of fundamental topics in Earth surface science with application to natural resources, geohazards, geoscience-based engineering, and stewardship of the environment among others. EAR currently consists of two sections on Surface Earth Processes (SEP) and Deep Earth Processes (DEP), as well as other programs and initiatives that include Education Programs, Special Programs, Critical Zone Observatories (CZOs), and Paleo-Perspectives on Climate Change.[1] The SEP section, with its programs of Geobiology and Low-Temperature Geochemistry; Geomorphology and Land Use Dynamics; Hydrologic Sciences; Sedimentary Geology and Paleobiology; and Education currently provides the primary support in the nation for research relevant to Earth surface processes, although Earth surface process-related research is only one portion of the total research portfolio that SEP and its programmatic disciplines are tasked to address.

Importantly, Earth surface processes is a rapidly progressing, integrative research field that also overlaps other disciplinary elements within EAR, the other divisions in GEO, and other NSF directorates including Biological Sciences; Computer and Information Science and Engineering; Engineering, Mathematical and Physical Sciences; and Social, Behavioral, and Economic Sciences. For interdisciplinary research projects that may pertain to one or more funding units, NSF program officers may share proposals among units to consider joint funding. A recent report of the Committee of Visitors commended program officers in SEP for their efforts to identify opportunities for co-funding from both within and outside NSF (NSF, 2008). Although this committee encourages continuation of that type of discretionary sharing of proposals among programs and sections, this informal option is not considered sufficient by itself to foster and maintain growth in the field of Earth surface processes. Comments received by the committee during the course of this study confirm this view. Rather, explicit interdisciplinary programs and funding opportunities are more effective for such work and are especially important in the initial stages of developing broad, interdisciplinary initiatives.

NSF has existing mechanisms to accommodate research that may not fall naturally under one existing disciplinary research program. These mechanisms could allow interdisciplinary research initiatives, such as those suggested by this committee for Earth surface processes, to take root and develop:

- Support for workshops, symposia, and conferences funds proposals in special areas of science and engineering that bring together experts to discuss recent research or education findings or to expose other researchers or students to new research and education techniques. These activities could be supported by multiple units and may lead to new programs, for example, NSF MARGINS, that employ multidisciplinary approaches.[2] The Frontiers in the Critical Zone workshop[3] is another example that led to the development of Critical Zone Observatories (CZOs; see Box 2.5);

---

[1] EAR website, http://www.nsf.gov/div/index.jsp?div=EAR [accessed August 8, 2008].
[2] http://www.nsf.gov/funding/pgm_summ.jsp?pims_id=13516&org=EAR&from=home [accessed October 5, 2008].
[3] http://www.czen.org/content/frontiers-exploration-critical-zone.

- Intradirectorate initiatives, for example, CZOs;
- Definition of priority areas and special grant opportunities (for example, the research area "Biocomplexity in the Environment" at NSF began as a special competition in 1999 and grew in the intervening years to become an established multidirectorate program of "Dynamics of Coupled Natural and Human Systems" in 2007); the Collaboration in Mathematical Geosciences Program[4] is an example of a special funding opportunity (through 2010) to enable crossdisciplinary research and education at the intersection of mathematical sciences and geosciences;
- Dear Colleague Letters announce special funding opportunities for interdisciplinary research. For example, three Dear Colleague Letters issued in February 2009 are relevant to Earth surface processes: (1) Environment, Society, and the Economy (ESE) seeks to increase collaboration between the geosciences and the social and behavioral sciences in integrated studies related to environment, society, and economics;[5] (2) Emerging Topics in Biogeochemical Cycles (ETBC) bridges the biological, atmospheric, geological, oceanographic, and hydrological sciences through topics related to biogeochemical cycles and processes;[6] (3) Multiscale Modeling (MSM) enhances support for research that links the biological sciences with the Earth system sciences in the area of multiscale modeling;[7]
- Crosscutting activities—funding programs in which two or more NSF directorates and/or other federal agencies participate; for example, the Dynamics of Coupled and Natural Human Systems Program (CNH),[8] through coordination by the Directorate of Biological Sciences, the Directorate of Geosciences, and the Directorate of Social, Behavioral, and Economic Sciences, promotes interdisciplinary analyses of human and natural system processes and interactions (the U.S. Forest Service recently began participation as a partner in the conduct of annual CNH competitions);
- NSF-wide activities (activities in which all NSF directorates participate, for example, "Carbon and Water in the Earth System"[9]);
- Science and Technology Centers, for example, the National Center for Earth-surface Dynamics (NCED), which carry out sustained, interdisciplinary research over 5 to 10 years (see Box 2.7); and
- Grants for rapid response research (RAPID) and Early concept Grants for Exploratory Research (EAGER)[10] (these provide relatively modest amounts, respectively,

---

[4] http://www.nsf.gov/pubs/2009/nsf09520/nsf09520.htm?org=NSF.
[5] http://www.nsf.gov/pubs/2009/nsf09031/nsf09031.jsp.
[6] http://www.nsf.gov/pubs/2009/nsf09030/nsf09030.jsp.
[7] http://www.nsf.gov/pubs/2009/nsf09032/nsf09032.jsp
[8] http://www.nsf.gov/funding/pgm_summ.jsp?pims_id=13681&org=EAR&from=home [accessed October 5, 2008].
[9] http://www.nsf.gov/funding/pgm_summ.jsp?pims_id=13651&org=OCE&from=home [accessed October 5, 2008].
[10] http://www.nsf.gov/pubs/policydocs/pappguide/nsf09_1/gpg091print.pdf [accessed March 19, 2009].

for quick-response research on natural or anthropogenic disasters and unanticipated events, and for high-risk, exploratory, and potentially transformative research; interdisciplinary projects could be supported by multiple relevant programs).

It is important to note that, in addition to NSF, federal research activities related to Earth surface processes exist within the U.S. Geological Survey (USGS), National Aeronautics and Space Administration, and U.S. Department of Agriculture, among other agencies (see Appendix D). Although NSF is the appropriate focal point for providing broad support for research on Earth surface processes, these activities at other agencies present significant opportunities for coordination and partnership.

4.2 OTHER PARTNERSHIPS

Although the majority of funding for academic Earth surface research has come from the federal government, a consistent record with important future potential exists to develop supporting partnerships with other groups. States have long supported locally targeted surface process research, especially as related to water quality, and state water quality standards often play a dominant role in landscape restoration projects. Earth surface processes are also important to a diverse array of industries, from environmental consulting and forestry to hydrocarbons and mining. Forest products companies have supported research related to problems of erosion and sedimentation in logged upland areas, for example. Scientists working on depositional systems are in a somewhat unusual position in that seismic and other data on subsurface structure are expensive to obtain and hence are held mostly by private industry. Given that petroleum companies also work on fundamental and wide-ranging problems in predicting surface evolution over long time scales, areas of common research interest between academia and industry are readily identified and partnerships between these groups of researchers could be actively pursued.

Furthermore, although the descriptions of the supporting mechanisms for Earth surface processes research in this chapter focus on implementation within the U.S. domestic research framework, a clear need exists for international collaborations and networks of research that stretch across key global surface environments. The types of problems addressed by research in this field by their very nature often require a global outlook or approach. Although support for research partnerships across international boundaries is difficult to garner, federal agencies may play some role in facilitating international scientific exchanges. The committee is aware of large research efforts in Earth surface processes in China, Australia, and Europe that could serve as initial contacts for such exchanges and later development of full research partnerships. Such international research may pay off handsomely in the future in terms of acquired knowledge as well as predictive capability to mitigate or avoid future problems.

4.3 DATA COLLECTION AND DISTRIBUTION, MODELING, TOOL DEVELOPMENT, AND COMMUNITY RESEARCH FACILITIES AND SITES

The rapid growth of quantitative research in Earth surface processes has been fueled in part by the large quantity and high quality of data collected during the past decade from satellites, airborne lidar (light detection and ranging), seismic reflection, geochronologic and geochemical instruments, and other technologies (see also Chapters 1 and 2), and the increased availability and power of new computing and modeling techniques to use these data. Maintaining existing instruments and supporting the development of new technologies to measure processes at Earth's surface are necessary parts of this fabric of growth and continued advancement of the field. Community laboratories and field sites also facilitate links among different technological applications and communication among disciplines.

Access to Data and Metadata

Broad access to data and metadata is necessary to advance any field of research but is especially important for interdisciplinary fields where barriers to communication among researchers are more likely to exist. Existing programs to enable data access include those supported by federal agencies as parts of their missions (see Appendix D), as well as the development of new cyber-databases. For example, EarthChem, an NSF-funded program, provides databases on rock and sediment chemistry (http://www.Earthchem.org/EarthchemWeb/index.jsp). The CZOs (Box 2.5) and the Critical Zone Exploration Network have enabled a group of Critical Zone scientists to compile data for sites that span gradients in environmental variables (for example, lithology and climate) to drive the development of models and understanding with respect to regolith evolution. These researchers are establishing protocols and a template for input and sharing of data generated by CZO research projects. The NCED (Box 2.7) has established a web-based system by which NCED and other community data are shared freely. The National Center for Airborne Laser Mapping (NCALM) provides research-grade airborne laser swath data to the national research community (see also Chapter 1), and those collected for the GeoEarthScope program are available at the GEON OpenTopography portal.[11] The National Oceanographic and Atmospheric Administration's National Geophysical Data Center[12] is an example of a readily accessible online resource for geophysical data that includes observations from space and models of the seafloor and solid Earth. The results produced by large observatory networks such as the Long Term Ecological Research (LTER) Network also are made available through the Internet via links from the network web page to databases maintained by network member sites (Box 4.1). Other examples are efforts to track diseases (for example,

---

[11] http://facility.unavco.org/project_support/es/geoEarthscope/.
[12] http://www.ngdc.noaa.gov/.

BOX 4.1
Data Sharing in the LTER Network

The LTER Network is a long-running program with much experience that can inform the Earth surface community of its success in promoting data sharing among LTER scientists and with the broader science community. A data access policy was formalized by the LTER Network Coordinating Community in 1997 (http://lternet.edu/data/netpolicy.html). Significant latitude exists in the way in which the policies are implemented by individual sites, but all share several common attributes:

Two types of datasets exist: The first type is freely accessible within two to three years of collection with minimal restrictions, whereas the second type is available only with written permission of the principal investigator or researcher.

- Type 2 datasets are rare.
- Data are available to the scientific community in a timely way.
- Investigators contributing data to LTER databases receive appropriate acknowledgment for use of the data by other researchers.
- Documentation of the dataset is adequate for the data to be used by researchers not involved in its original collection.
- Data must continue to be available even if an investigator leaves the project or NSF no longer funds the LTER site.
- Adequate quality assurance and control are maintained.
- Datasets may not be sold or distributed by the recipient.
- Investigators have reasonable opportunity to have first use of data they collect.
- Each site's data management policy and data accessibility are peer-reviewed every three years during an NSF-required formal site review process.

Development of functional data-sharing capability was a major challenge for the network. In the 1980s, LTER scientists were no more likely to share data than other scientists. Attempts to make data available online were met with resistance by many investigators even though the benefits of data sharing were widely recognized. Consequently, a committee of site data managers was convened to investigate ways to promote data sharing and yet protect the data collected by individual investigators from use without attribution. By 1990, network guidelines had been established requiring each site to develop its own data management policy. The benefits of establishing guidelines rather than a policy were that researchers whose data would be part of the data-sharing effort were engaged in developing policies that worked for them and multiple approaches emerged. Approaches that proved to be effective were adopted by other sites and refined so that by 1993 each site had its own data-sharing policies in place. The 1997 LTER data-sharing policy is the product of the LTER Network's efforts to make needed data available to researchers worldwide. Currently more than 4,000 datasets from 26 LTER sites are listed in the online data catalogue (http://metacat.lternet.edu/knb/). Importantly, renewal of funding is contingent upon adherence to network data management standards.

cholera, severe acute respiratory syndrome [SARS], avian influenza - H5N1 virus), global climate change, and the Integrated Ocean Observing System. Analysis of these types of frequently used data sharing arrangements reveals common attributes such as openness, transparency, quality control of data, operational flexibility, respect for intellectual property, management accountability, dedicated infrastructure, and long-term financial support.

Despite general community agreement that data-sharing and maintenance of large datasets are important, large-scale data sharing is not easily implemented and requires concerted action by the research community and sponsors. Consensus policies lend confidence to the process of the eventual treatment, acknowledgment, documentation, and distribution of the data (see Box 4.1).

Modeling

Until recently, there was no coordinated effort among Earth surface processes researchers to build community models. Instead, individual research groups developed models to support their research, and commonly these models were not readily available to others. Even if made available, the constantly changing language in which the models were written and the lack of interface instructions made using these models difficult. Notable exceptions include those where individuals or organizations made considerable effort to make the models accessible and usable by others.[13] Recently, NSF supported the formation of the Community Surface Dynamics Modeling System (CSDMS) to provide access to software generated by the community, create linked dynamic models for landscape basin evolution, and foster collaboration and community sharing of model and model development (see also Section 2.3). Fostering partnerships between Earth surface process software development and other relevant geoscience computation initiatives (e.g., the Computation Infrastructure for Geodynamics [CIG]) and the modeling efforts of the geospatial sciences (e.g., the National Center for Geographic Information Analysis) also could enable emerging technologies in other fields to be used for Earth surface process modeling. These are important steps and could lead to a framework to build the many models anticipated to emerge from each of the four research initiatives.

Technology Development

New technologies have greatly increased the quantity and flexibility of data collection on many aspects of the Earth's surface at scales from individual mineral surface sites to continents (see Chapter 2, as well as Boxes 1.2, 2.1, 2.4, 2.6). The importance of continued

---

[13] For example, the U.S. Army Corp of Engineers Hydrologic Engineering Center programs for flow, flood, and sediment transport; the Excel-based programs by Gary Parker (University of Illinois) for a range of channel hydraulics and sediment transport calculations; and the USGS-supported Multi-Dimensional Surface Water Modeling System.

support for *existing* technologies such as satellite sensors, lidar and InSAR (interferometric synthetic aperture radar), geochronologic methods, geophysical instruments, and biogeochemical tools is emphasized throughout this report (see also Appendix C). Of particular importance has been the development of numerous geochronologic methods that support the dating of different materials and time scales (Boxes 1.2 and 2.4). Community laboratory facilities such as the NSF-supported Purdue Rare Isotope Measurement [PRIME] laboratory at Purdue University and the Lawrence Livermore National Laboratory's Center for Accelerator Mass Spectrometry (CAMS) have provided relatively inexpensive access for researchers and are especially important in this regard.

New technologies and improvement of existing ones for measuring different properties, forms, fluxes, and rates are greatly needed, and these technologies should be readily available to the community. Some current areas ripe for support and advance are summarized here. Earth's near-surface environment is largely inaccessible but for drill holes or trenches, which provide only limited information about subsurface characteristics that dictate runoff, moisture dynamics and geochemical weathering, and geomechanical strength properties. As described in Section 2.5, geophysical methods such as electrical resistance tomography, self-potential, ground-penetrating radar, seismic refraction, and neutron probe surveys have been little used in Earth surface process research but may provide key observational data obtainable by no other means. Improvements in data analysis, the availability of and knowledge about instruments, and the development of other methods are needed, and support for community-level efforts in this area could provide significant advances to improve data density and software analysis. Airborne hyperspectral surveys, integrated with lidar data are beginning to be used to map vegetation structure and topography. High-resolution quantification of form via lidar (both airborne and ground-based) is revolutionizing field work and analysis of many surface processes.

We are entering an era of being able to measure real time Earth surface processes with high accuracy. Hundreds of wireless devices uplinked to broadband systems can be positioned across a landscape and monitored simultaneously to document sap flow, soil moisture, overland flow, groundwater levels, air temperature, solar radiation, suspended sediment concentrations, and other attributes. Already, a few examples of such observations are radically improving our understanding of process dynamics. Research is under way to develop software to manage these data systems. NSF has a Science and Technology Center devoted to wireless networks, the Center for Embedded Networked Sensing. Support for further development of instruments and software will usher in this "real-time" era. Other instruments that still are being prototyped for application to Earth surface and other research include terrestrial and marine autonomous vehicles for collecting environmental data, particle and organism tracking with various types of radio-frequency devices, and instruments for measuring physical properties at the base of ice sheets.

Community Research Facilities and Sites

In recent years we have seen a rebirth of the use of both large-scale experimental facilities and intensive field monitoring campaigns in the study of Earth surface processes. This rebirth is associated with the widely recognized need for facilities (experimental or field observatories) where community-level, interdisciplinary collaborations can tackle fundamental problems. Such collaborations are enabled by new technologies of observation and new models to both guide and test observations, and are motivated by urgent new questions.

Many individual researchers have built experimental facilities for particular research projects, but the maintenance of these by individual scientists is taxing and facilities often fall into decline once a project is completed. National facilities enable concentration of shared resources so that technically challenging problems can be undertaken in a cost-effective manner. Such facilities also create opportunities or even the necessity for interdisciplinary research. In the United States, just one national experimental facility at the Saint Anthony Falls Laboratory (NCED, Box 2.7) exists for Earth surface process studies. Not only does it house a large number of flumes, but adjacent to the facility is the Outdoor StreamLab in which field-scale channels with riparian zones are built and a wide range of ecogeomorphology studies are conducted. As with the Biosphere 2[14] indoor experimental hillslopes, building large, experimental landforms at the finer end of such features found in nature enables greatly improved measurements, represents more realistic biotic processes, and reduces or eliminates many issues about upscaling.

The simultaneous call from many disciplines for relatively large-scale field "observatories," which was initiated many years ago, is now being met (CZOs [Box 2.5], National Ecological Observatory Network [NEON], LTER sites [Box 4.1], and planning for hydrological observatories). Various experimental watersheds operated by universities and federal agencies (for example, the Walnut Gulch Experimental Watershed,[15] and the Big Spring Run, Pennsylvania, floodplain-wetland restoration experiment funded by the Pennsylvania Department of Environmental Protection, the USGS, and the U.S. Environmental Protection Agency) provide opportunities for research at field sites where long-term monitoring of environmental variables occurs (see also Section 2.6). These sites are essential to advancing the research initiatives and addressing the grand challenges identified in this report. Given the range and diversity of science questions and ecogeomorphic regions, continuation and growth of various observatories is appropriate.

A very different kind of facility, one focused on data synthesis, has been successfully developed in ecology at the National Center for Ecological Analysis and Synthesis

---

[14] http://www.b2science.org/index.html.
[15] http://www.tucson.ars.ag.gov/unit/gis/wg.html.

(NCEAS).[16] Synthetic activities are supported by the Consortium of Universities for the Advancement of Hydrologic Science, Inc. (CUAHSI) for water-related research. CSDMS and NCED organize workshops, but there is no synthesis center in the form of NCEAS in Earth surface processes. Such a center, perhaps in response to emerging observations from the various observatories, could prove valuable.

4.4 DEVELOPING THE HIGH-PRIORITY RESEARCH INITIATIVES

The four high-priority initiatives described in Chapter 3 cut across research disciplines and themes and draw together the interacting aspects of the natural and social sciences needed to investigate the Earth's surface. The four initiatives require sustained, interdisciplinary, intellectual and technical collaboration that includes linked field, laboratory, and modeling efforts. Coordinated, interdisciplinary field campaigns ("joint field campaigns") and the use of shared laboratory facilities, observatories and experimental field sites, research centers, federal data repositories, and other organized networks and programs can help build the communication base necessary to link research in many fields. Exchange of information and results and mutual education through organized workshops, symposia, or wireless networking media are also a part of the process of community building to develop the four research initiatives and foster growth in the field.

The four initiatives are major undertakings that will require significant, coordinated effort to mobilize and direct. This would be in addition to continued support and building of existing disciplinary research programs related to study of the Earth's surface at NSF, particularly those programs within the SEP section of EAR. Although nascent interactions among various disciplinary groups have begun, challenges to effective and fulfilling collaboration include bridging cultural differences across disciplines. Nonetheless, the success of some of the aforementioned interdisciplinary centers shows that the broad research community has evolved to the point of being able to overcome these barriers. Some potential organizational mechanisms, as they could be applied explicitly to these four initiatives, are given below for consideration by NSF and the science community. Because the four initiatives are strongly synergistic, the development of mechanisms to enable fluid, ongoing sharing of insight and information among all four initiatives may further encourage their success.

Interacting Landscapes and Climate

Fostering collaboration between Earth surface scientists and atmospheric scientists, with targeted research options specifically for young investigators, could be centerpieces to

---

[16] http://www.nceas.ucsb.edu/.

achieve the objectives outlined in this initiative (see Chapter 3). Significant progress could be made in facilitating effective collaborations through the following:

- Interdisciplinary training and education through community workshops and climate-land surface summer schools for graduate students, postdoctoral researchers, and faculty (for example, the National Center for Atmospheric Research's Cooperative Program for Operational Meteorology, Education and Training [COMET] program[17] or the NSF Integrative Graduate Education and Research Traineeship programs).
- Joint field campaigns tackling shared fundamental problems of the interaction of climate and Earth surface processes that could involve atmospheric scientists and Earth surface scientists.
- Modeling collaborations between climate scientists and Earth surface science researchers that could include, for example, regional and microclimate modeling (including land-cover effects), orographic precipitation, hydrological responses, wind and wave energy, glacier dynamics and sea level, and landscape evolution (the CSDMS could play an active role).
- Organized participation by joint engineering and Earth surface science teams in the development of satellite and land-based sensors for monitoring and quantifying environmental factors and processes relating to climatic control of Earth surface processes. The requirements from these sensors may include rainfall, evapotranspiration, streamflow, soil temperature and moisture content, storm surge and wave energy, near-surface winds, sediment transport, solute transport, and glacier sliding velocities and mass balance.

Quantitative Reconstruction of Landscape Dynamics Across Time Scales

Addressing the objectives in this initiative requires focus on (1) application and development of specific analytical techniques and tools, including cosmogenic isotopic analyses, low-temperature and detrital thermochronology, isotopic measurements, molecular biological analyses, lidar mapping of landforms, and three-dimensional imaging of subsurface structures and (2) effective combination of these analyses and their results in coupled models. As discussed in Chapter 3, the necessary intellectual collaborations include deep- and surface-Earth sciences, with a balance between scientists from industry and those from academia. Significant progress could be made in facilitating effective collaborations through the following:

---

[17] http://www.comet.ucar.edu/outreach/index.htm.

- Development of natural deep-time laboratories to focus on reconstruction of Earth's surface evolution from short-term to geologic time scales—from instants to eons—with emphasis on using quantitative methods to identify thresholds and abrupt changes and to study the interplay of tectonics, climate, biota, and surface processes. This effort would include a targeted program of intensive drilling and seismic acquisition.
- Targeted projects with joint industry and academic participation, organization, and products to apply noncommercially sensitive portions of three-dimensional seismic surveys along with associated core data, potentially acquired through new shallow drilling endeavors, to key initiative objectives.
- Continued development of cosmogenic, optically stimulated luminescence, isotopic, and low-temperature thermochronological methods for quantifying the rates of past processes and the timing of abrupt changes; encouraging the use of existing community laboratories (PRIME laboratory, for example) and research on application of these tools to new minerals could strengthen these collaborations.
- Coordinated community development of fully coupled climate-tectonic-geochemical, ecological-surface process models that engages existing numerical modeling initiatives (for example, CSDMS, CIG, and other geochemical initiatives) and organizes targeted interdisciplinary workshops with participation from the atmospheric sciences and from ecological and paleontological disciplines.
- Development and support of shared, community experimental laboratory facilities for landscape research across time scales and testing models in a controlled environment.

Coevolution of Ecosystems and Landscapes

The recent emergence of ecohydrology, geobiology, and ecogeomorphology to describe current research activities shows the growing effort to work at the interface of Earth and biological sciences. The initiative proposed here would focus on taking this interest much farther and deeper. The ultimate goal is to create opportunities for discoveries that are equally advanced in the fields of ecology and Earth surface processes and are obtainable only because of strong interdisciplinary interactions (see also Chapter 3). Specific mechanisms to develop these collaborations include the following:

- Establishing working groups that organize regular meetings to focus on research at the interface of ecosystems and landscape processes and evolution. Special sessions on ecosystems and landscapes at national meetings of many organizations (for example, the American Geophysical Union, Geological Society of America, Association of American Geographers, and Ecological Society of America) are a

start. The workshop-style meetings envisioned here could complement these types of special sessions by providing an opportunity for interdisciplinary teaching and for formulating joint research plans.
- A community-level modeling program in which ecologists and Earth scientists collaborate on contributions to models for short-term forecasts and predictions on long time-scales of the coevolution of ecosystems and landscapes. For community-level modeling, collaboration is essential to avoid unrealistic, overly simplistic treatment of coupled ecosystem and landscape processes.
- Joint field campaigns conducted by climate scientists and Earth surface scientists, including ecologists, geomorphologists, and hydrologists, which will enable underlying mechanisms to be quantified that link biota, ecosystems, and Earth surface processes. New coalitions are necessary to build reliable models for prediction on the shorter-term and evolutionary time scales.
- The network of observatory sites (CZOs, NEON, LTER, and possibly hydrologic sites) can be used explicitly to explore ecological and Earth surface processes. Future observatory site selection could be optimized for both ecologic and physical science objectives.
- Codevelopment of instrumentation, geochemical, geophysical, and geochronological tools could facilitate significant advances. For example, shallow geophysical surveys should provide information important to ecosystem processes, and biologically mediated isotopic traces can reveal hydrologic pathways and residence times.

The Future of Landscapes in the "Anthropocene"

The overarching goal for this initiative is to transform our understanding of integrated human-landscape systems characteristic of the "Anthropocene" and improve predictive capabilities of how they might evolve in the future. Advancing the objectives for this initiative requires collaboration among climate scientists, Earth surface scientists, and ecologists, among others, and a range of social and behavioral scientists including economists, political scientists, and human geographers. Transformational advances require bridging the real and perceived gaps between the approaches of natural and social scientists. Collaboration with geospatial scientists, engineers, and applied practitioners is also needed. Success of an initiative to address the future of human-landscape systems includes effective transfer of scientific knowledge toward societal application. Increasing the awareness of the general public (at community research facilities) of the role of human activities and the sustainable management of Earth's systems is also necessary. Workshops and working groups are central to the initial coordination and collaboration required among the disciplinary fields. Specific mechanisms to develop these collaborations include the following:

- Workshops to bring together Earth surface scientists and social scientists to build integrated community approaches, research questions, methodologies, scales of inquiry, and theories for human-landscape systems. Successful, sustained collaborations will allow study of complex mutual interactions between societies and Earth surface systems and will allow response to the challenge of incorporating human behavior in mechanistic models.
- Workshops that engage geospatial scientists with Earth surface scientists to examine the integration of existing and emerging geospatial technologies at specific experimental field sites and to process and synthesize remote-sensing data. These data serve as inputs to model development that could eventually extend from local, controlled experimental sites to global models.
- Development of community field and modeling centers to acquire the data necessary for new integrative and predictive models that involve multiple stressors within human-dominated landscapes, including social processes that influence those interactions.
- Focused field studies in sensitive environments that are determined to be most vulnerable to anthropogenic change, including coastal and urban areas where human populations are concentrated, mountain and polar environments where melting glaciers translate into water resource issues and hazards, and arid and semiarid areas that are increasingly affected by drought and variability in streamflow associated with climate change and expanding urban populations. These studies could take advantage of existing environmental observatories, such as CZOs and LTERs, and develop mechanisms for broad synthesis, as exemplified by the NCEAS.
- Collaborative research using engineered landscapes and restoration and redesign projects could provide relatively controlled conditions, including those of time and rate. Such research could improve the fundamental knowledge of processes relevant to a range of environments and problems, such as testing hypotheses about building self-maintaining ecogeomorphic systems. This collaboration could involve engineers and applied practitioners working with Earth surface scientists, which also could serve as a means to transfer scientific knowledge toward societal application.

References

Barras, J., S. Beville, D. Britsch, S. Hartley, S. Hawes, J. Johnston, P. Kemp, Q. Kinler, A. Martucci, J. Porthouse, D. Reed, K. Roy, S. Sapkota, and J. Suhayda. 2003. Historical and projected coastal Louisiana land changes: 1978-2050. USGS Open File Report 03-334, 39 pp. (revised January 2004).

Bernhardt, E.S., M.A. Palmer, J.D. Allan, G. Alexander, K. Barnas, S. Brooks, J. Carr, S. Clayton, C. Dahm, J. Follstad-Shah, D. Galat, S. Gloss, P. Goodwin, D. Hart, B. Hassett, R. Jenkinson, S. Katz, H.M. Kondolf, P.S. Lake, R. Lave, J.L. Meyer, T.K. O'Donnell, L. Pagano, B. Powell, and E. Sudduth. 2005. Synthesizing U.S. river restoration efforts. *Science* 308:636-637.

Bertoni, C., and J. Cartwright. 2005. 3D seismic analysis of slope-confined canyons from the Plio-Pleistocene of the Ebro continental margin (western Mediterranean). *Basin Research* 17:43-62.

Brantley, S.L., T.S. White, A.F. White, D. Sparks, D. Richter, K. Pregitzer, L. Derry, J. Chorover, O. Chadwick, R. April, S. Anderson, and R. Amundson. 2006. Frontiers in Exploration of the Critical Zone: Report of a workshop sponsored by the National Science Foundation (NSF), October 24-26, 2005, Newark, De. Available at http://www.czen.org/content/frontiers-exploration-critical-zone (accessed June 10, 2009).

Cox, J.D. 2005. *Climate Crash: Abrupt Climate Change and What It Means for Our Future*. Washington, D.C.: The National Academies Press, 224 pp.

Cox, R., Zentner, D.B., Rakotondrazafy, A.F.M., and Rasoazanamparany, C.F. 2010. Shakedown in Madagascar: Occurrence of lavakas (erosional gullies) associated with seismic activity. *Geology* 38: in press.

Crutzen, P.J., and E.F. Stoermer. 2000. The "Anthropocene." *IGBP Newsletter* 41(1):17-18.

Fagherazzi, S., M. Marani, and L.K. Blum. 2004. *The Ecogeomorphology of Tidal Marshes*. Washington, D.C.: American Geophysical Union.

Fimmen, R.L., D.deB. Richter, D. Vasudevan, M.A. Williams, and L.T. West. 2007. Rhizogenic Fe-C redox cycling: A hypothetical biogeochemical mechanism that drives crustal weathering in upland soils. *Biogeochemistry* 87:127-141. DOI 10.1007/s10533-007-9172-5.

Foley, J.A., R. DeFries, G.P. Asner, C. Barford, G. Bonan, S.R. Carpenter, F.S. Chapin, M.T. Coe, G.C. Daily, H.K. Gibbs, J.H. Helkowski, T. Holloway, E.A. Howard, C.J. Kucharik, C. Monfreda, J.A. Patz, I.C. Prentice, N. Ramankutty, and P.K. Snyder. 2005. Global consequences of land use. *Science* 309(5734):570-574.

Frankel, K.L., K.S. Brantley, J.F. Dolan, R.C. Finkel, R.E. Klinger, J.R. Knott, M.N. Machette, L.A. Owen, F.M. Phillips, J.L. Slate, and B.P. Wernicke. 2007. Cosmogenic ^{10}Be and ^{36}Cl geochronology of offset alluvial fans along the northern Death Valley fault zone: Implications for transient strain in the eastern California shear zone. *Journal of Geophysical Research—Solid Earth* 112:B06407.

Galewsky, J., and G.H. Roe. 2008. Climate over landscapes: The emerging links between geomorphology and the atmospheric sciences. Available at http://earthweb.ess.washington.edu/roe/Publications/GalewskyRoe_whitepaper.pdf (accessed May 26, 2009).

Grootes, P.M., and M. Stuiver. 1997. Oxygen 18/16 variability in Greenland snow and ice with 10^{-3}- to 10^5-year time resolution. *Journal of Geophysical Research* 102(C12): 26455-26470.

Hilley, G.E., and J.R. Arrowsmith. 2008. Geomorphic response to uplift along the Dragon's Back Pressure Ridge, Carrizo Plain, California. *Geology* 36:367-370.

Ivanov, V.Y., R.L. Bras, and E.R. Vivoni. 2008. Vegetation-hydrology dynamics in complex terrain of semiarid areas: I. A mechanistic approach to modeling dynamic feedbacks. *Water Resources Research* 44:W03429.

Iverson, N. 2008. Challenges and opportunities in Earth surface processes: A glaciocentric perspective. Powerpoint presentation to Committee on Challenges and Opportunities in Earth Surface Processes, June 24. On file at the National Academies Public Access Records Office.

Jibson, R.W. 2005. Landslide hazards at La Conchita, California. U.S. Geological Survey Open-File Report 2005-1067, 12 pp.

Koons, P.O. 1990. The two-sided orogen: Collision and erosion from the sand box to the Southern Alps, New Zealand. *Geology* 18:679-682.

Major, J.J., J.E. O'Connor, G.E. Grant, K.R. Spicer, H.M. Bragg, A. Rhode, D.Q. Tanner, C.W. Anderson, and J.R. Wallick. 2008. Initial fluvial response to the removal of Oregon's Marmot Dam. *Eos* 89(27):241.

Merritts, D.M., G. Hilley, J.R. Arrowsmith, W. Carter, W. Dietrich, J. Jacobs, S. Martel, J. Roering, R. Shrestha, and N.P. Snyder. 2009. Workshop on Studying Earth Surface Processes with High-Resolution Topographic Data, Boulder, Colo. June 15-18, 2008. Available at http://lidar.asu.edu/FinalReport_NSF_Whitepaper_NCALMworkshop_June2008ForDistribution.pdf (accessed May 20, 2009).

Mitchell, N.C., and J.M. Huthnance. 2007. Comparing the smooth, parabolic shapes of interfluves in continental slopes to predictions of diffusion transport models. *Marine Geology* 236(3-4):189-208.

Nagel, J., T. Dietz, and J. Broadbent. 2009. Workshop on Sociological Perspectives on Global Climate Change (summary of workshop proceedings held at NSF in Arlington, Va. May 30-31, 2008).

National Research Council (NRC). 2001a. *Basic Research Opportunities in Earth Science*. Washington, D.C.: National Academy Press, 154 pp.

NRC. 2001b. *Compensating for Wetland Losses Under the Clean Water Act*. Washington, D.C.: National Academy Press, 348 pp.

NRC. 2004. *Facilitating Interdisciplinary Research*. Washington, D.C.: The National Academies Press, 332 pp.

NRC. 2007a. *Elevation Data for Floodplain Mapping*. Washington, D.C.: The National Academies Press, 168 pp.

NRC. 2007b. *Evaluating Progress of the U.S. Climate Change Science Program: Methods and Preliminary Results*. Washington, D.C.: The National Academies Press, 178 pp.

NRC. 2008. *Hydrology, Ecology, and Fishes of the Klamath River Basin*. Washington, D.C.: The National Academies Press, 272 pp.

National Science Foundation (NSF). 2008. Surface Earth processes: Earth meets life, air and water. Powerpoint presentation to American Geophysical Union Surface Earth Processes Town Hall Meeting, December 18. On file at the National Academies Public Access Records Office.

Owen, L.A., G. Thackray, R.S. Anderson, J. Briner, D. Kaufman, G.H. Roe, W. Pfeffer, and C. Yi. 2008. Integrated research on mountain glaciers: Current status, priorities and future prospects. *Geomorphology* 103(2):158-171.

Perron, J.T., J.W. Kirchner, and W.E. Dietrich. 2009. Formation of evenly spaced ridges and valleys. *Nature* 460:502-505. doi:10.1038/nature08174.

Pfirman, S., and the AC-ERE. 2003. Complex environmental systems: Synthesis for Earth, life, and society in the 21st century, A report summarizing a 10-year outlook in environmental research and education for the National Science Foundation, 68 pp. Available at http://www.nsf.gov/geo/ere/ereweb/acere_synthesis_rpt.cfm (accessed June 1, 2009).

Robinson, D.A., A. Binley, N. Crook, F.D. Day-Lewis, T.P.A. Ferre, V.J.S. Grauch, R. Knight, M. Knoll, V. Lakshmi, R. Miller, J. Nyquist, L. Pellerin, K. Singha, and L. Slater. 2008. Advancing process-based watershed hydrological research using near-surface geophysics: A vision for, and review of, electrical and magnetic geophysical methods. *Hydrological Processes* 22(18):3604-3635.

Roering, J.J. 2008. How well can hillslope evolution models "explain" topography? Simulating soil transport and production with high-resolution topographic data. *GSA Bulletin* 120(9/10):1248-1262.

Sposito, G. 2008. *The Chemistry of Soils*. New York: Oxford University Press, 329 pp.

Steffen, W., P.J. Crutzen, and J.R. McNeill. 2007. The Anthropocene: Are humans now overwhelming the great forces of nature? *Ambio* 36:614-621.

Syvitski, J.P.M., C.J. Vörösmarty, A.J. Kettner, and P. Green. 2005. Impact of humans on the flux of terrestrial sediment to the global coastal ocean. *Science* 308(5720):376-380.

Titus, J.G. (ed). 1988. *Greenhouse Effect, Sea Level Rise, and Coastal Wetlands*. Washington, D.C.: U.S. Environmental Protection Agency.

Tuchman, B.W. 1978. *A Distant Mirror: The Calamitous 14th Century*. New York: Random House.

Tucker, G.E., and R.L. Slingerland. 1994. Erosional dynamics, flexural isostasy, and long-lived escarpments. *Journal of Geophysical Research* 99(B6):12229-12243.

United Nations, Department of Economic and Social Affairs, Population Division. 2009. World Population Prospects: The 2008 Revision, Highlights, Working Paper No. ESA/P/WP.210.

Vajjhala, S., A. Krupnick, E. McCormick, M. Grove, P. McDowell, C. Redman, L. Shabman, M. Small. 2007. Rising to the Challenge: Integrating Social Sciences into NSF Environmental Observatories. (A report by Resources for the Future for NSF; available at http://www.rff.org/RFF/Documents/NSFFinalReport.pdf; accessed 2 July 2009).

Vorosmarty, C.J., M. Meybeck, B. Fekete, K. Sharma, P. Green, and J. Syvitski. 2003. Anthropogenic sediment retention: Major global-scale impact from the population of registered impoundments. *Global and Planetary Change* 39:169-190.

Walter, R.C., and D.J. Merritts. 2008. Natural streams and the legacy of water-powered mills. *Science* 319:299-304.

Whipple, K.X. 2009. The influence of climate on the tectonic evolution of mountain belts. *Nature Geoscience* 2:97-104.

Willett, S.D. 1999. Orogeny and orography: The effects of erosion on the structure of mountain belts. *Journal of Geophysical Research* 104(B12):28957-28981.

Zalasiewicz, J., M. Williams, A. Smith, T.L. Barry, A.L. Coe, P.R. Bown, P. Brenchley, D. Cantrill, A. Gale, P. Gibbard, F.J. Gregory, M.W. Hounslow, A.C. Kerr, P. Pearson, R. Knox, J. Powell, C. Waters, J. Marshall, M. Oates, P. Rawson, and P. Stone. 2008. Are we now living in the Anthropocene? *GSA Today* 18:4-8.

Appendixes

APPENDIX A

Biographical Sketches of Committee Members and Staff

COMMITTEE MEMBERS

Dorothy J. Merritts, *Chair,* is a professor in the Department of Earth and Environment at Franklin & Marshall College in Lancaster, Pennsylvania. In 2004-2005 she was the Flora Stone Mather Visiting Distinguished Professor at Case Western Reserve University in Cleveland, Ohio. She has expertise in streams, rivers, and other landforms and the impact of humans and geologic processes on landscape evolution. In the western United States, she conducted pioneering research on the San Andreas fault of coastal California, and her international work focuses on fault movements in South Korea, Indonesia, Australia, and Costa Rica. Her primary research in the eastern United States is on streams in the mid-Atlantic Piedmont, particularly in southeastern Pennsylvania and northern Maryland, where she is investigating the impact on streams of the transformation of woodland and wetland forests to a predominantly agricultural and mixed industrial-urban landscape since European settlement. She is the author of two textbooks and more than 40 scientific articles and the editor and contributing writer for numerous scientific books. Dr. Merritts has done extensive work on inquiry-based learning in the classroom, particularly for non-science majors at the undergraduate level, and has assisted in presenting original inquiry-based materials and demonstrations online through the Science Education Resource Center at Carleton College, Minnesota. Dr. Merritts received her B.Sc. in geology from Indiana University of Pennsylvania, her M.Sc. in engineering geology from Stanford University, and her Ph.D. in geology from the University of Arizona.

Linda K. Blum is a research associate professor in the Department of Environmental Sciences at the University of Virginia. Dr. Blum's current research focuses on how microorganisms bring about geomorphologic changes in salt marshes. Her research includes a long-term interest in the linkage between microbial community spatial patterns and processes and microbe-plant interactions. Dr. Blum was previously the chair of the National Research Council (NRC) Panel to Review the Department of Interior's Critical Ecosystem Studies Initiative, a member of the Committee on Restoration of the Greater Everglades Ecosystem, and a previous and current member of the Committee on Independent Review of Everglades Restoration Progress. She earned a B.S. and an M.S. in forestry from Michigan Technological University and a Ph.D. in soil science and microbial ecology from Cornell University.

Susan L. Brantley has served on the faculty of the Department of Geosciences at the Pennsylvania State University since 1986. She is currently a full professor. She is also director of the Earth and Environmental Systems Institute and Center for Environmental Kinetics Analysis at the Pennsylvania State University. Her research interests focus on the chemical, physical, and biological processes associated with the circulation of aqueous fluids in shallow hydrogeologic settings. She has published more than 110 papers that have discussed aspects of water-rock-biota interaction, the kinetics of dissolution and precipitation of minerals in the laboratory and in the field, surface chemistry of minerals, environmental water problems, biogeochemical cycles, volcano-water interactions, soil chemistry, and water interactions in metamorphic rocks. She has been awarded a David and Lucile Packard Fellowship, was a National Science Foundation (NSF) Presidential Young Investigator, and was a fellow of the AGU (American Geophysical Union). She has served on several NRC committees including, most recently, her stint as vice chair of the Panel on Solid-Earth Hazards, Resources, and Dynamics, established to write the solid-Earth contribution to *Earth Science Applications from Space: A Community Assessment and Strategy for the Future* (Decadal Study). She has also served on the advisory committee for the Directorate of Geosciences at NSF (2003-2005), during which time she served on the Committee of Visitors to review the Division of Earth Sciences (EAR) Instrumentation and Facilities Program (2004) and chaired the Committee of Visitors to review the NSF EAR Section on Surface Earth Processes (2005). She received her A.B. in chemistry and her M.A. and Ph.D. in geological and geophysical sciences from Princeton University.

Anne Chin is courtesy professor in the Department of Geography of the University of Oregon. Previously, she was associate professor of geography at Texas A&M University. She is a fluvial geomorphologist with research interests in the energetics of mountain, dryland, and urban rivers. She seeks to understand how landscapes interact with a range of human impacts over diverse spatial and temporal scales. Her work also addresses landscape management and restoration. This line of inquiry explores the interplay between physical and biological processes on the one hand and the social forces that shape policy on the other. Dr. Chin was the recipient of the 2004 Grove Karl Gilbert Award for Excellence in Geomorphological Research from the Association of American Geographers. Her work has appeared in a range of journals in the geosciences, including the *American Journal of*

Science, Geophysical Research Letters, Geomorphology, Journal of the American Water Resources Association, Journal of Geology, Annals of the Association of American Geographers, Progress in Physical Geography, Environmental Management, and *BioScience*. She has served on numerous professional advisory boards and is past chair of the Geomorphology Specialty Group of the Association of American Geographers. In 2006-2007, she was visiting scientist and director of the Geography and Regional Science Program of the National Science Foundation. Dr. Chin holds a B.A. from the University of California, Los Angeles, and a Ph.D. from Arizona State University, both in geography.

William E. Dietrich (NAS) is a professor in the Department of Earth and Planetary Science at the University of California, Berkeley. He also has an appointment in the Department of Geography and the Earth Sciences Division of the Lawrence Berkeley National Laboratory and is affiliated with the Archeological Research Facility. He is co-founder of the National Center for Airborne Laser Mapping and a member of the National Center for Earth-surface Dynamics. His Berkeley-based research group is focusing on mechanistic, quantitative understanding of the form and evolution of landscapes and the linkages between ecological and geomorphic processes. He has numerous publications and honors, including being named a member of the National Academy of Sciences and a fellow of the American Academy of Arts and Sciences, both in 2003. Dr. Dietrich received his B.A. from Occidental College and his M.S. and Ph.D. from the University of Washington.

Thomas Dunne (NAS) is a professor in the Bren School of Environmental Science and Management at the University of California, Santa Barbara. He is a hydrologist and a geomorphologist, with research interests that include field and theoretical studies of drainage basin and hillslope evolution, sediment transport and floodplain sedimentation, and sediment budgets of drainage basins. He served as a member of the NRC Committees on U.S. Geological Survey Water Resources Research; Opportunities in the Hydrologic Sciences; Alluvial Fan Flooding; and Future Roles, Challenges, and Opportunities for the U.S. Geological Survey. He was elected to the National Academy of Sciences in 1988 and the American Academy of Arts and Sciences in 1993. He has acted as a scientific adviser to the United Nations; the governments of Brazil, Taiwan, Kenya, Washington, and Oregon; and several U.S. federal agencies. He is a recipient of the American Geophysical Union Horton Award. Dr. Dunne holds a B.A. from Cambridge University and a Ph.D. in geography from the Johns Hopkins University.

Todd A. Ehlers is a professor and chair of geology and geodynamics at the Institute for Geoscience, Universität Tübingen, Germany. Prior to this he was an associate professor and associate chair of the Department of Geological Sciences at the University of Michigan. Dr. Ehlers's research interests are in the topographic evolution of mountains over geologic

APPENDIX A

time scales. Techniques used in his research include numerical modeling of geomorphic, geodynamic, and atmospheric processes; low-temperature thermochronology; and cosmogenic radionuclides. His research group is active in the areas of quantifying the glacial erosion histories of mountains and the evolution of climate, topography, and deformation in the Himalaya and Andes mountains. He is an associate editor for *Tectonics*, coeditor-in-chief for *Earth Science Reviews*, and a fellow of the Geological Society of America. Dr. Ehlers received a B.A. from Calvin College and M.S. degrees in geophysics and geology and a doctorate degree in geophysics from the University of Utah.

Rong Fu is a professor at the Jackson School of Geosciences at the University of Texas at Austin. Dr. Fu's research aims at understanding the role of the atmospheric hydrological cycle in determining the stability of the Earth's climate. She uses satellite and in situ observations and numerical models to identify the mechanisms that control this interaction between water cycle and surface climate, and uses them to explain natural variability and anthropogenic forced changes in rainfall, cloudiness, and water vapor distribution. In recent years, her research has been focused on the coupling between rainfall, rainforest, and biomass burning in the Amazon and on convective transport of water vapor in the tropics and over the Tibetan Plateau. Dr. Fu has served on a variety of national and international panels and programs, including the NRC review panels for the National Aeronautics and Space Administration (NASA) Carbon Cycle Science Program, Cloud and Aerosol Program, the National Oceanic and Atmospheric Administration (NOAA) and NSF panels for Climate Prediction and Drought Research, and the panels for the U.S. and International CLIVAR. She is also a long-term member of the NASA Aura, SesWinds and UARS Science Teams. Dr. Fu received her B.S. from the Department of Geophysics at Peking University and her M.A. and Ph.D in atmospheric sciences from Columbia University.

Christopher Paola is a professor in the Department of Geology and Geophysics at the University of Minnesota at Minneapolis, where he has been since 1983. He worked for the U.S. Geological Survey (USGS) Cascades Volcano Observatory from 1988 to 1990 and has held visiting appointments at the University of Genoa (Italy) and Columbia University. From 2003 through early 2008 he served as director of the National Center for Earth-surface Dynamics, a National Science Foundation Science and Technology Center devoted to transdisciplinary research on the evolution and behavior of the Earth's surface. He has published more than 60 papers on surface dynamics and stratigraphy. He is a fellow of the Geological Society of America and the American Geophysical Union. Dr. Paola graduated from Lehigh University with a B.S. in environmental geology and an M.Sc. in applied sedimentology from the University of Reading (U.K.). After receiving an Sc.D. in marine geology from the Woods Hole Oceanographic Institution-Massachusetts Institute of Technology Joint Program in Oceanography in 1983, he joined the faculty at the University of Minnesota.

Kelin X. Whipple is a professor at the School of Earth and Space Exploration at Arizona State University. Previously, he was a professor in the Department of Earth, Atmospheric, and Planetary Sciences at the Massachusetts Institute of Technology (MIT). Dr. Whipple's research focuses on the interaction of climate, tectonics, and surface processes in the sculpting of the Earth's surface; mechanics of river incision into bedrock; dynamics of channel and sedimentation processes on alluvial fans; and experimental and field study of debris-flow rheologies. His current research activities focus on the geomorphic evolution of fluvial bedrock channel and alpine glacial valley systems. His active projects and interests span a range from small-scale modeling and investigation of the physics of bedrock channel erosion; to reach-scale modeling of the dynamics of bedrock channel evolution; to neotectonic studies of active deformation using geomorphic tools; to quantitative investigation of linkages between tectonics, climate, and surface processes at mountain range scale. Dr. Whipple is editor of *AGU Editor's Choice: Surface Processes* and associate editor of the *Journal of Geophysical Research—Earth Surface*; he has served as associate editor of the *Journal of Geophysical Research—Solid Earth*. He is chair of the American Geophysical Union's Sediment and Landscape Dynamics Committee and sits on the AGU Erosion and Sedimentation Committee. Dr. Whipple received his Ph.D. and M.S. in geology from the University of Washington and his B.A. in geology from the University of California, Berkeley.

NATIONAL RESEARCH COUNCIL STAFF

Elizabeth A. Eide, senior program officer, is a geologist with specialization in isotope geochronology applied to crustal processes. Prior to joining the National Research Council, she was a research scientist and team leader at the Geological Survey of Norway and managed the survey's $^{40}Ar/^{39}Ar$ geochronology laboratory. She received her Ph.D. in geology from Stanford University and a B.A. in geology from Franklin & Marshall College.

Jared P. Eno was a research associate with the Board on Earth Sciences and Resources (BESR). Before coming to the National Academies, he interned at Human Rights Watch's Arms Division, working on the 2004 edition of the *Landmine Monitor Report*. Jared received his A.B. in physics from Brown University.

Courtney R. Gibbs is a program associate with the Board on Earth Sciences and Resources. She received her degree in graphic design from the Pittsburgh Technical Institute in 2000 and began working for the National Academies in 2004. Prior to her work with BESR,

APPENDIX A

Ms. Gibbs supported the Nuclear and Radiation Studies Board and the former Board on Radiation Effects Research.

Nicholas D. Rogers is a financial and research associate with the Board on Earth Sciences and Resources. He received a B.A. in history, with a focus on the history of science and early American history, from Western Connecticut State University in 2004. He began working for the National Academies in 2006 and has primarily supported BESR on a broad array of Earth resource, mapping, and geographical science issues.

APPENDIX B

Community Input

Because of the highly interdisciplinary nature of Earth surface processes research, the committee sought input from a broad cross section of the research community (Table B.1). Solicited input was obtained in one of two ways: (1) presentations from and discussion with external panelists representing disciplines in Earth surface process research; and (2) responses to a nationally and internationally distributed questionnaire containing three questions developed by the committee (Box B.1). Unsolicited input was also accepted from individuals or groups of researchers who were aware of the study and chose to submit written comments or documents for the committee's consideration. The committee also used input from white papers that were to be drafted after various National Science Foundation (NSF)-sponsored workshops conducted during 2007-2008 in areas specifically oriented toward Earth surface process research.

A list of speakers who presented at the committee's meetings and the committee's questionnaire are included at the end of this appendix. The questionnaire was forwarded and published in 25 Earth science-oriented listservs, websites, newsletters, journals, and professional societies (Section B.1); 8 boards, study committees, and sections within the National Academies; and 9 offices or programs within the National Aeronautics and Space Administration (NASA), National Oceanographic and Atmospheric Administration (NOAA), U.S. Department of Agriculture (USDA) Forest Service, and U.S. Geological Survey (USGS). As of August 1, 2008, there were a total of 83 responses to the questionnaire. Of these, 55 self-identified their sector as "university faculty," 14 as "public sector," 8 as "private sector," 2 as "other," 2 as "graduate student," and 2 as "NGO" (nongovernmental organization). Responses came from at least 25 states and 4 foreign countries (note that not all respondents provided their name and affiliation).

Responses to each question covered a broad range, but a few themes emerged repeatedly. The broadest area of agreement among respondents was the significance of new remote-sensing technologies for the field. Dating techniques, modeling, computing technology, and

APPENDIX B

data access were also highlighted by many comments. Conceptual advances in our understanding of climate change and the role and significance of human factors, as well as partnerships between different fields were called out as important steps by many respondents as well. The two most common challenges listed by respondents were funding issues and barriers to interdisciplinary collaboration.

TABLE B.1 Organizations and Their Associated Publications or Listservs to Which the Questionnaire Announcement Was Sent

Organization, Publication, or Listserv
Association of American Geographers (AAG) Newsletter and selected specialty groups
American Geological Institute (AGI) GeoSpectrum, Government Affairs Monthly Review
American Geophysical Union (AGU) *Eos*
American Society of Limnology and Oceanography listserv
America Society for Photogrammetry and Remote Sensing (ASPRS) listserv
Association for Women Geoscientists listserv
Canadian Geomorphology Research Group listserv
Clay Minerals Society listserv
European Science Foundation
Friends of Mineralogy listserv
GSA Connection
Gilbert Club listserv
Gulf Coast Association of Geological Societies listserv
International Association of Geomorphologists listserv
International Association of Hydrogeologists listserv
Mineralogical Society of America listserv
National Speleological Society listserv
Paleontological Society listserv
Seismological Society of America listserv
Society for Sedimentary Geology listserv
Society of Economic Geologists listserv
Society of Exploration Geophysicists listserv
Society of Vertebrate Paleontology listserv
Soil Science Society of America listserv
U.S. Permafrost Association listserv

> **BOX B.1**
> **Committee Questionnaire**
>
> At the request of the National Science Foundation, the National Research Council is conducting a study that will assess (1) the state of the art of the multidisciplinary field of Earth surface processes, (2) the fundamental research questions in the future for the field, and (3) the challenges and opportunities facing the research community and the nation in answering these questions. The committee is addressing the task by considering research on the dynamic biological, chemical, physical, and human processes, interactions, and feedback mechanisms that affect the shape of Earth's surface across a range of spatial and temporal scales, from continental interiors to the oceans, and from polar to equatorial regions. The committee is dedicated to generating a report that will be used by a wide audience including policy makers, agency managers, scientists from many disciplines, and society.
>
> The report will have the greatest impact on future research if it has strong input from a broad spectrum of the interested community. During its few scheduled study meetings, the committee cannot hear from all of the many interested individuals who have important input to this topic, so the committee seeks your help in the form of written contributions to the following set of questions:
>
> 1. What have been the four most significant conceptual and/or technological advances in Earth surface processes in the last 15 years?
> 2. What are two emergent and fundamental questions that Earth surface processes research can address?
> 3. What challenges (organizational, administrative, conceptual, philosophical, etc.) exist in conducting the research needed to answer the fundamental questions identified in Question 2?

B.2 LIST OF SPEAKERS AT COMMITTEE MEETINGS

Teofilo Abrajano, NSF
Rafael L. Bras, Massachusetts Institute of Technology
Oliver Chadwick, University of California, Santa Barbara
Terry Chapin, University of Alaska
Michael Church, University of British Columbia
Louis Derry, Cornell University
Martin Doyle, University of North Carolina
Tom Drake, Office of Naval Research
Michael Ellis, NSF
Jon Foley, University of Wisconsin
Christian France-Lanord, CNRS-Nancy, France

APPENDIX B

Joseph Galewsky, University of New Mexico
Arthur Goldstein, University of New England
Neal Iverson, Iowa State University
Matthew Larsen, USGS
Randy McBride, George Mason University
Gregory Okin, University of California, Los Angeles
Denise Reed, University of New Orleans
Robert Stallard, USGS
Brad Werner, Scripps Institution of Oceanography
James Whitcomb, NSF

APPENDIX C

Observing and Measuring Earth Surface Processes

Scientific research is based on the collection of reliable and accurate data and observations from which interpretations, models, and concepts can be generated and tested. Some of the conceptual advances, opportunities, and challenges in Earth surface processes research highlighted in this report have been made possible by the development, particularly within the last 10 to 15 years, of a number of analytical tools to observe and measure features or activities at the Earth's surface. This appendix reviews prominent tools, instruments, and techniques used in Earth surface processes research as a basic reference; many of these tools were highlighted by community experts during the committee's open meetings and by respondents to the committee's community questionnaire. The tools here are grouped very generally into those that address (1) measurement and visualization of Earth's surface including topography; (2) timing or age of landforms or Earth surface processes; (3) composition (chemical or biological) of Earth's surface features; and (4) mass flux or physical properties of the landscape.

C.1 REMOTE SENSING

Widespread availability of digital data of Earth's topography and other surface attributes, collected predominantly by satellites and aircraft, allows study of vast regions, comparison of different parts of Earth's surface, monitoring of Earth surface hazards in real time, and quantification of properties such as terrain dissection, vegetation type and height, ground moisture content, and soil mineralogy. These data can be viewed in the form of imagery that has become familiar to many through various Internet resources. The combination of Earth imagery and accurate topographic data is particularly powerful. Widely used tools include the following:

- Light detection and ranging (lidar) uses a laser beam mounted on an aircraft, satellite, truck, ship, or hand-held device to measure accurate distances to a target

point on land or under-water that is accurately located in a geographic coordinate system. The data can be used to produce a digital surface model (DSM), which can then be processed to yield digital elevation models (DEMs), digital terrain models, contours, and three-dimensional feature data.
- Interferometric synthetic aperture radar (InSAR) uses radar mounted on an airplane or satellite to measure the strength and round-trip travel time of a microwave signal to gather elevation data; these data can be used to produce DSMs and DEMs, among other products, as with lidar.
- High-resolution swath bathymetry uses a transducer mounted on a ship to measure the round-trip travel time of sonar waves emitted in the water below the ship. The time elapsed between emission and reception of the signals allows the depth to the seafloor and other features to be determined.
- Sensors mounted on satellite platforms measure various bands of the electromagnetic spectrum to produce images of land and water surfaces at wavelengths appropriate to determine vegetation type, general chemical composition, temperature, water content, and other aspects of a given land or water surface. Data from federally, commercially, or internationally owned and operated satellites are available either freely online or for purchase. The reader is referred to NRC (2008) for information on a variety of such satellite missions.
- Global positioning system (GPS) technology measures a highly accurate three-dimensional position on Earth's surface at a precise time relative to known positions of several satellites. Information can be used to determine rates of tectonic motion, including uplift rates. Differential GPS measurements have yielded detailed observations of glacier surface velocities and temporal variations in those velocities that are important in understanding the glacier dynamics.

C.2 DATING TECHNIQUES AND CHRONOLOGY

The issue of "timing" or "age" of landforms and surface attributes is of great interest to Earth surface processes research. Accurate age data enable the history and rates of processes to be determined—for example, dating techniques are being used to study the rates of floodplain sedimentation and soil processes, faulting histories and paleoseismic recurrence intervals, and the topographic evolution of mountain ranges. A wide range of dating techniques exists; some of the most frequently used types are listed below:

- Fallout-derived and short-lived isotopes are used for quantifying sediment transport, deposition, and chronology on time scales of days (for example, beryllium, lead, and cesium isotopes).

- Thermoluminescence and cosmogenic radionuclide (CRN) techniques can be used for dating terrestrial deposits that range in age from several thousand to millions of years. These techniques may include, for example, the use of beryllium, aluminum, or chlorine isotopes or optically stimulated luminescence (OSL).
- Low-temperature thermochronometer techniques are used to provide information on hundred thousand- to million-year rates of erosion and topographic evolution (for example, U-Th-Sm/He, fission track, and $^{40}Ar/^{39}Ar$ thermochronology techniques).
- Sidereal methods count annual events, such as dendrochronology (tree ring dating) and varve chronology (sediment layer dating).

C.3 BIOGEOCHEMISTRY

Although some biogeochemical data can be gleaned from remote-sensing applications, most biogeochemical measurements require physically sampling rock, water, or vegetation material and subsequently conducting analysis in a laboratory. Such data are therefore slow to accumulate compared to remotely sensed data and are sparser with respect to distribution over time and space. However, this sparseness is balanced by the rich nature of biogeochemical data that can provide information about more than 3,000 minerals, more than 100 chemical elements, and many chemical species and isotopes.

Biogeochemical tools are used to study weathering rates and mechanisms, issues of soil fertility, interpretation of biological fluxes, contaminant geochemistry, bioavailability and mobility of elements, soil organic matter, fractions of carbon, evapotranspiration, elemental cycling, exposure age dating, interpretation of hydrology, and biotic characteristics, among others. Broadly speaking, these tools can be categorized by the type of analysis they support. Primary types of analyses are listed below, along with examples of strategies for measurement.

- Elemental analysis—emission and absorption spectroscopies and X-ray and photo-electron analysis can produce bulk elemental compositions, as well as spot analysis as a function of depth or position.
- Chemical speciation analysis—chromatography, electrochemistry, and X-ray absorption analysis can be used to identify the chemical species present in solids, liquids, gases, and biological material.
- Mineralogical analysis—X-ray diffraction, thermogravimetric analysis, electron and neutron diffraction, and electron microscopy can identify the mineralogical structures and particle sizes of solid phases.
- Textural analysis—electron and optical microscopes, surface topographic analysis, image analysis, particle size analysis, and neutron diffraction can obtain particle

size and grain size and shape, as well as support topographic analysis and pore size analysis.
- Isotopic analysis—mass spectrometry coupled with various inlet strategies can identify the isotopic composition of phases, spots, layers, individual molecules, or biota in solids, liquids, or gases.
- Organic carbon analysis—mass spectrometry and coulometry can identify soil organic matter, fractions of carbon, fertility of soil, and other information.
- Molecular biological analysis—polymerase chain reaction, metagenomics, and some mass spectrometric tools can yield data to identify genomes, DNA, RNA, enzymes, and metabolisms.
- Surface area analysis—Brunauer-Emmet-Teller (BET) measurements, porosimetry, and neutron diffraction can determine surface areas, particularly as they relate to transfer of elements into water.
- Dissolved organic matter (DOM) analysis—fluorescence spectroscopy provides information on the chemical composition (DOM) of marine water, freshwater, and wastewater.

C.4 PHYSICAL PROPERTIES

Obtaining quantitative physical information about Earth's surface is possible with tools that allow measurement of the dynamics and kinematics of mass flux and storage and of material properties such as density, temperature, magnetism, porosity, and permeability. Methods used to obtain this type of information include the following:

- Advanced seismic methods that use two- and three-dimensional imaging with acoustic waves to resolve fossilized surface features to resolutions of 1 meter or better in the field and, up to several kilometers below the surface;
- Electromagnetic resistivity, tomography, and magnetometers measure electrical and/or magnetic fields to yield data on groundwater flow and subsurface geometry and structures; electrodes or magnetometers are placed at the surface or in boreholes;
- Rock core-logging tools are attached to a drillstem to measure downhole neutron density, natural gamma radiation, resistivity, and sonic properties of rock during drilling to yield data that indicate density, porosity, permeability, and to some degree, composition of the rock column;
- Ground-penetrating radar yields two- and three-dimensional subsurface images at imaging depths between about 1 and 50 meters; radar pulses are emitted and collected from units that can be hand-held or mounted on vehicles;

Appendix C

- Stream gauges are part of a national network coordinated by the U.S. Geological Survey and partners in 50 states to measure flow and/or water level in streams and rivers; and
- Miniaturized instruments and dataloggers including pressure transducers monitor, for example, flooding occurrence in floodplains or storm surges along coastlines, surface heave, and electrical conductivity.

C.5 REFERENCE

National Research Council (NRC). 2008. *Earth Observations from Space: The First 50 Years of Scientific Achievements.* Washington, D.C.: The National Academies Press.

APPENDIX D

Achievements in Earth Surface Processes

As described in this report, Earth surface processes is a new field that has emerged over the past two decades as a result of growing recognition of the need for interdisciplinary science able to answer questions that do not reside squarely within the realm of a single discipline. The 1992 Chapman Conference on Tectonics and Topography and the 2003 Penrose Conference on climate, tectonics, and landscape evolution were landmark events that both illustrated this new approach and indicated that thresholds had been crossed in terms of scientific interest. Growth of the field has also been reflected in a variety of community and organizational developments, including formation of Earth surface processes-related research groups and laboratories; creation of new interdisciplinary Earth surface processes majors at universities and colleges; establishment of multiinstitute-multiresearcher centers or observatories (Boxes 2.5 and 2.7); publication of textbooks, book series,[1] and new journals dedicated to research in the field;[2] and establishment of focus groups by international professional organizations (e.g., the European Surface Processes Group;[3] the American Geophysical Union (AGU) Earth and Planetary Surface Processes focus group, established in 2008).

College and university hiring patterns also point to increasing interest in and recognition of the field. Analysis of new tenure-track faculty position advertisements in *Eos*, a widely read weekly magazine published by the AGU that has a very broad readership across all of the Earth, atmosphere, and ocean sciences, suggests an increasing trend in Earth surface processes positions. Figure D.1 shows the number of advertisements placed in *Eos* for the years 1990, 1994, 2001, and 2007 that were considered to be for positions

---

[1] http://www.elsevier.com/wps/find/bookdescription.cws_home/BS_DESP/description#description.

[2] For example, the journals *Earth Surface Processes* and *Landforms* (a publication of the British Society for Geomorphology), the *Journal of Geophysical Research—Earth Surface* (a publication of the American Geophysical Union), and the journal *Geomorphology* each seek to publish academic articles specifically related to study of Earth's surface.

[3] http://rock.esc.cam.ac.uk/eusurfaceprocesses/.

APPENDIX D

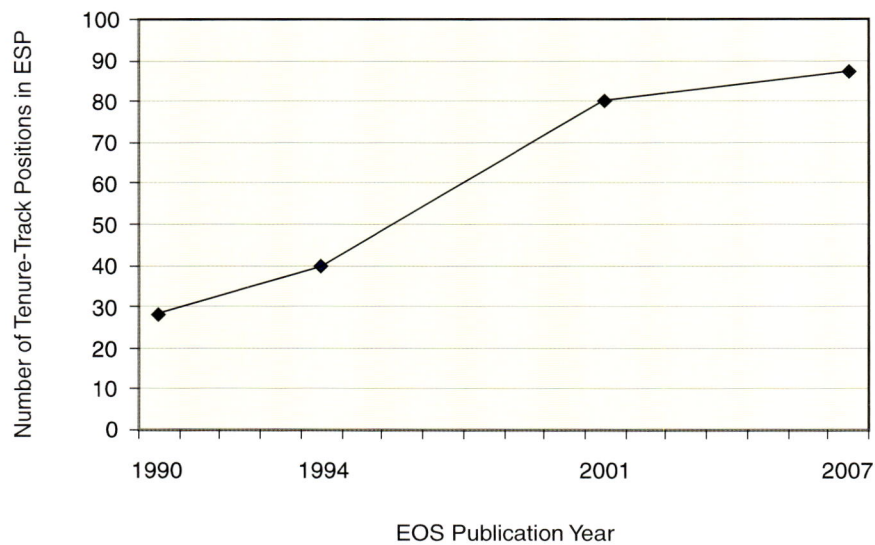

FIGURE D.1 A tally of the number of tenure-track positions in Earth surface processes (ESP on the vertical axis) advertised in *Eos* in selected years suggests a significant increase in the number of these positions between 1990 and 2007. In the absence of the specific term "Earth surface processes" or "surficial processes" in the job description, a position was identified as Earth surface process-related if it (1) focused on processes operating at the surface of the Earth and/or (2) described an intent to integrate research and teaching about Earth's surface across two or more disciplines, departments, or schools.

in Earth surface processes. Categorizing past position postings using a term that has only recently come into use is a difficult and, to a degree, subjective task because institutions only occasionally labeled their open positions specifically for "Earth surface processes." Interdisciplinary science such as Earth surface processes can be troublesome to fit into the framework of departments that generally reflect traditional boundaries between disciplines. Thus, positions were categorized as Earth surface process positions if the advertisement described a position that was (1) focused on processes operating at the surface of the Earth and/or (2) intended to integrate across disciplines.

A similar survey of articles in the prominent monthly journal *Geology*, published by the Geological Society of America (GSA), shows a steady increase in the percentage of articles on Earth surface processes (Figure D.2). This journal encourages publication of articles from across all of the Earth sciences, so it is also widely accessible to and read by a diverse spectrum of researchers. As encountered in the examination of *Eos*, the categorization of journal articles as Earth surface process-related was not unique because few articles used the full term Earth surface processes to title or describe their work. However, the same criteria used

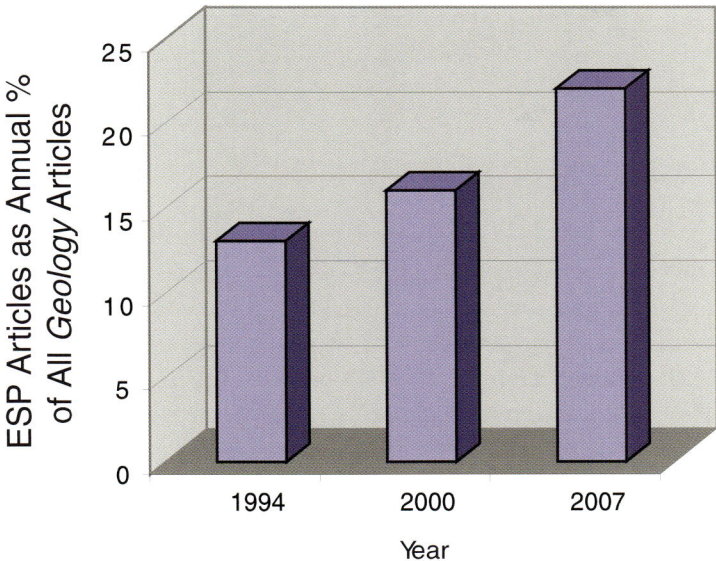

FIGURE D.2 Percentage of all articles in selected years of GSA's *Geology* journal that were considered to be on topics within Earth surface processes.

for the *Eos* review of tenure-track positions were employed to determine if an article was directly related to the field of Earth surface processes.

Efforts in science education and outreach also indicate increasing recognition of Earth surface processes. In addition to Earth science education initiatives such as the National Science Foundation (NSF)-supported Earth science literacy initiative[4] and educational programs of the AGU[5] and GSA,[6] education projects on Earth surface processes have engaged professional educators, professional societies, local organizations (such as museums), communications experts, and members of the research community. Examples of these endeavors include Earth surface process-related exhibits and features at museums across the country, such as the National Center for Earth-surface Dynamics (NCED)-University of Minnesota-Science Museum of Minnesota collaboration (Box 2.7) and the Smithsonian Museum of Natural History exhibition "Dig It! The Secrets of Soil."[7] These efforts have also helped increase participation in Earth surface processes by students from under-represented groups.

---

[4] http://www.Earthscienceliteracy.org/.
[5] http://www.agu.org/sci_soc/education/.
[6] http://www.geosociety.org/educate/.
[7] http://forces.si.edu/soils/.

APPENDIX D

Nonetheless, a great deal of work remains to be done to capitalize on the familiarity and intrinsic appeal of landscapes to attract students from underrepresented groups into the sciences and engineering.

D.1 FEDERAL SUPPORT FOR EARTH SURFACE PROCESSES

Numerous federal agencies provide support for research and other activities in Earth surface processes. In addition to NSF (see Chapter 4), the U.S. Geological Survey (USGS), National Aeronautics and Space Administration (NASA), U.S. Department of Agriculture (USDA), Environmental Protection Agency, National Oceanic and Atmospheric Administration (NOAA), Federal Emergency Management Agency, U.S. Army Corps of Engineers, Naval Research Laboratories, and others have activities that research, use, analyze, regulate, monitor, or manage aspects of the United States' continental and marine landscapes and the processes that shape them. The next sections briefly examine examples of research programs and activities at the USGS, NASA, and USDA that relate directly to Earth surface processes.

U.S. Geological Survey

The USGS conducts formal and informal interdisciplinary research activities related to Earth surface processes on topics that include geohazards; land use and land change and effects on the carbon cycle; origin and development of land features; and coastal research on erosion, deposition, and extreme weather and their effects on wetlands, ecosystems, and beaches (Larsen, 2008; USGS, 2008). The USGS is also the only federal agency that the committee could identify with designated Earth Surface Processes teams of scientists that dedicate part of their activities to Earth surface-related research and a program called "Earth Surface Dynamics," which examines climate-landscape oriented issues under the aegis of USGS Climate Change Science activities. Modest funding for all of these activities is derived from various programs within the USGS, as well as through reimbursable funds from other federal, state, and local government agencies.

The USGS also supports and coordinates large, publicly available datasets and operates critical monitoring networks that contribute to research in Earth surface processes and other fields. The Earth Resources and Observation Science (EROS) Center manages, develops, and conducts research on aerial, satellite, land-use or land-cover, elevation, map, and other data and images.[8] The U.S. stream gauging program[9] provides continuous data to support sound management of the water resources in the nation's streams and rivers. Notably, a

---

[8] http://edc.usgs.gov/ [accessed October 5, 2008].
[9] http://water.usgs.gov/wid/html/SG.html#HDR0 [accessed October 6, 2008].

decrease in funds to support the stream gauge monitoring network in the last decade has resulted in significant data loss and decreased capability to contribute to this management effort (Church, 2008; USGS, 2008).

National Aeronautics and Space Administration

The ability of satellites to measure features of the Earth's surface globally at high resolution has established an increasing role for NASA in Earth surface processes research. Through the Earth Observing System, satellites produce publicly available data for long-term global observations of the land surface, biosphere, solid Earth, atmosphere, and oceans.[10] These instruments and data contribute to a range of research projects to understand, manage, and predict various Earth surface processes (Appendix C). NASA is also involved in the development of new sensors and technologies to record Earth surface data (NRC, 2007a).

In addition to designating a portion of its own funding to developing sensors and collecting and analyzing Earth observing data, NASA has cooperative arrangements with other agencies (NOAA and the USGS, for example) to transfer some sensors from research to operational or application-oriented functions (NRC, 2007b). NASA also solicits proposals from the academic research community to conduct work that employs Earth observing data and research results toward practical societal benefits.[11] Of the currently operating NASA, NOAA, or USGS satellite missions for making Earth observations, many are past or nearing the end of their expected lifetimes. The report *Earth Science Applications from Space: National Imperatives for the Next Decade and Beyond* (NRC, 2007a) described a minimum of 17 high-priority missions that should be considered for development by the federal government to maintain and improve existing Earth observing capabilities over the coming decades; most of these proposed missions have applications directly relevant to the study of Earth's surface.

U.S. Department of Agriculture

The USDA mission includes six strategic goals, one of which is to "protect and enhance the nation's natural resource base and environment."[12] Fulfillment of that strategic goal includes active research and analysis of Earth's surface, including landforms, soils, forests, grasslands, wetlands, water, wildlife habitats, and energy resources. The Forest Service, Agricultural Research Service (ARS), and Natural Resource Conservation Service (NRCS) of the USDA perform many of these Earth surface data collection and analysis functions. National programs of the ARS, for example, include research in soil resource manage-

---

[10] http://eospso.gsfc.nasa.gov/ [accessed October 7, 2008].
[11] http://nasascience.nasa.gov/Earth-science/applied-sciences [accessed October 7, 2008].
[12] http://www.obpa.usda.gov/budsum/fy08budsum.pdf [accessed October 7, 2008].

ment, global change, water availability and watershed management, and pasture, forage, and rangeland systems.[13] The Forest Service, which manages public forests, grasslands, and ecosystems, conducts research and development projects in watershed science, landscape management, and soil research, among others, and provides cooperative support for grants awarded under the NSF program Coupled Natural and Human Systems (see also Chapter 4).

The USDA supports external research through cooperative research agreements and competitive grants and also supports various online resources to provide information, forecasts, and databases for the public on various agricultural topics.[14] Development of online mapping tools has allowed researchers to access various soil, agricultural, and land-use datasets in geographically referenced map coordinate systems.

D.2 REFERENCES

Church, M. 2008. Presentation to Committee on Challenges and Opportunities in Earth Surface Processes. Washington, D.C., March 17.

Larsen, M. 2008. Presentation to Committee on Challenges and Opportunities in Earth Surface Processes. Washington, D.C., March 17.

National Research Council (NRC). 2007a. *Earth Science Applications from Space: National Imperatives for the Next Decade and Beyond*. Washington, D.C.: The National Academies Press, 456 pp.

NRC, 2007b. *Assessment of the NASA Applied Sciences Program*. Washington, D.C.: The National Academies Press, 160 pp.

U.S. Geological Survey (USGS), 2008. Submitted to the committee, March 14. On file at the National Academies Public Access Records Office.

---

[13] http://www.ars.usda.gov/pandp/locations/NPSLocation.cfm?modecode=02-02-00-00 [accessed October 7, 2008].
[14] http://www.usda.gov/wps/portal/!ut/p/_s.7_0_A/7_0_1OB?navid=RESEARCH_RESOURCES&parentnav=RESEARCH_SCIENCE&navtype=RT [accessed October 7, 2008].

APPENDIX E

List of Acronyms

AGU	American Geophysical Union
ALSM	airborne laser swath mapping
AMS	accelerator mass spectrometry
ARS	Agricultural Research Service (USDA)
B.P.	before present
BROES	Basic Research Opportunities in Earth Science (NRC)
CAMS	Center for Accelerator Mass Spectrometry
CIG	Computation Infrastructure for Geodynamics
CNH	Dynamics of Coupled and Natural Human Systems Program
COMET	Cooperative Program for Operational Meteorology, Education and Training
CRN	cosmogenic radionuclide
CRONUS	Cosmic-Ray produced nuclide systematics on Earth
CSDMS	Community Surface Dynamics Modeling System
CUAHSI	Consortium of Universities for Advancement of Hydrologic Science, Inc.
CZEN	Critical Zone Exploration Network
CZO	Critical Zone Observatory
DEP	Deep Earth Processes
EAGER	Early-concept Grants for Exploratory Research (NSF)
EAR	Division of Earth Sciences (NSF)
EROS	Earth Resources and Observation Science Center (USGS)

ESE	Environment, Society, and the Economy
ETBC	Emerging Topics in Biogeochemical Cycles
GEO	Directorate for Geosciences (NSF)
GEON	Geosciences Network
GIS	geographic information systems
GSA	Geological Society of America
IRIS	Incorporated Research Institutions for Seismology
InSAR	interferometric synthetic aperture radar
Lidar	light detection and ranging
LTER	Long Term Ecological Research
MSM	Multiscale Modeling
MYRES	Meeting of Young Researchers in Earth Sciences
NASA	National Aeronautics and Space Administration
NAVSTAR	Navigation Satellite Timing and Ranging
NCALM	National Center for Airborne Laser Mapping
NCEAS	National Center for Ecological Analysis and Synthesis
NCED	National Center for Earth-surface Dynamics
NEON	National Ecological Observatory Network
NOAA	National Oceanic and Atmospheric Administration
NRC	National Research Council
NRCS	Natural Resource Conservation Service
NSF	National Science Foundation
OSL	optically stimulated luminescence
PRIME	Purdue Rare Isotope Measurement
RAPID	Grants for search (NSF)
SEP	Surface Earth Processes (NSF)
SMM	Science Museum of Minnesota
STC	Science and Technology Center (NSF)
UNAVCO	University NAVSTAR Consortium

Appendix E

UNESCO	United Nations Educational, Scientific, and Cultural Organization
USDA	U.S. Department of Agriculture
USGS	U.S. Geological Survey